Anthony

ELECTRO-OPTICAL
COMMUNICATIONS
DICTIONARY

ELECTRO-OPTICAL COMMUNICATIONS DICTIONARY

EDITED BY
DENNIS BODSON AND DAN BOTEZ

HAYDEN BOOK COMPANY, INC.
Rochelle Park, New Jersey

Acquisitions Editor: PRIJONO HARDJOWIROGO
Production Editor: MARSHALL E. OSTROW
Art Director: JIM BERNARD
Compositor: PUBLISHER'S PHOTOTYPE INTERNATIONAL, INC.
Printed and bound by: MAPLE-VAIL BOOK MANUFACTURING GROUP

Library of Congress Cataloging in Publication Data

Bodson, Dennis.
    Electro-optical communications dictionary.

    1. Electrooptics—Dictionaries. 2. Fiber optics—Dictionaries. 3. Telecommunica-
tion—Dictionaries.
I. Botez, Dan. II. Title.
TK5103.59.B63 1983   621.38′0414   83-4295
ISBN 0-8104-0961-5

1	2	3	4	5	6	7	8	9	PRINTING
83	84	85	86	87	88	89	90	91	YEAR

PREFACE

The purpose of this vocabulary is to promote communication, enhance understanding, and provide for precision in the terminology associated with fiber optic and lightwave communication system technologies. This vocabulary represents an initial attempt to bring together in one place a comprehensive listing of terms and definitions associated with these advancing technologies.

This vocabulary is intended for use by those who are or may become involved with fiber optic and lightwave communication systems; for example, those who design, develop, maintain, operate, use, administrate, manage, and manufacture communication equipment, data processing equipment, components, and systems. It should also prove useful to teachers, students, engineers, doctors, and other professional and technical personnel.

Approximately 2500 entries, with inversions and cross-references, are included. These terms and definitions are consistent with international, federal, industry, and technical society standards; for example, those of the International Organization of Standards, International Electrotechnical Commission, International Telegraph and Telephone Consultative Committee, and the Electronic Industries Association. It should be noted, however, that some definitions have not yet been formally standardized.

This compendium represents only the tip of the iceberg insofar as terminology in the emerging fields of fiber optics and lightwave communications is concerned. During the next few years, as lightwave communication systems are developed and installed, changes in technology will foster changes in terminology.

Dennis Bodson and Dan Botez

ABOUT THE EDITORS

Dennis Bodson was born in Washington, D.C., on July 7, 1939. He received B.E.E. and M.E.E. degrees from Catholic University, Washington, D.C., in 1961 and 1963, respectively, an M.P.A. degree from the University of Southern California's Washington Center for Public Affairs in 1976, and is currently pursuing his doctorate in electrical engineering.

As a Systems Engineer engaged in research and development, he served with the U.S. Army Materiel Command, Vitro Laboratories, and the Atlantic Research Corporation. He is currently on the staff of the National Communications Systems (NCS), Office of Technology and Standards, Washington, D.C., where his major responsibilities are the development of Federal Standards relating to telecommunications and fiber optic transmission systems. He serves on Standard Committees of the Electronic Industries Association, American Standards Institute, and the International Telegraph and Telephone Consultative Committee. He is the author of 12 scientific papers, some of which he presented at national and international conferences.

Mr. Bodson is a Registered Professional Engineer in the District of Columbia and the State of Virginia and is certified by the National Council of Engineering Examiners. He is also a senior member of the Institute of Electrical and Electronics Engineers and a Fellow of the Radio Club of America.

Dan Botez was born in Bucharest, Romania, in 1948. He received B.S. (summa cum laude), M.S., and Ph. D. degrees in electrical engineering from the University of California, Berkeley, in 1971, 1972, and 1976, respectively. His doctoral studies concerned characteristics of layers deposited over preferentially etched channels in gallium arsenide as well as novel optical devices made possible by this method. He continued his research in the semiconductor laser field as a Postdoctoral Fellow at the IBM T.J. Watson Research Center in Yorktown Heights, N. Y.

In 1977 he joined the staff at RCA Laboratories, Princeton, N.J., where his work led to the development of "thick window" high-radiance surface-emitting LEDs and the constricted double-heterostructure (CDH) laser, which represents one of the least temperature-sensitive commercially available diode lasers ever developed. Some of his more recent work has produced the CDH large-optical-cavity laser, which has allowed optical recording at one of the highest data rates ever achieved. In 1982 he was appointed a Research

Leader in RCA's Opto-Electronics Group. He has published over 50 scientific papers and has presented more than 30 invited and contributed papers at national and international conferences.

Dr. Botez received a 1979 RCA Outstanding Achievement Award for contributions to the development of a high-density optical recording system employing an injection laser. He is a member of Phi Beta Kappa and the Institute for Electrical and Electronics Engineers.

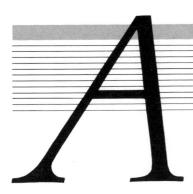

Å *Symbol for ANGSTROM*

abbe constant A mathematical expression for determining the correction for chromatic abberation of an optical system.
Note: It is usually expressed as

$$V = \frac{N_D - 1}{N_F - N_C}$$

or refractivity/dispersion where N_D, N_F, and N_C are the indices of refraction for light of the wavelengths of the D line of sodium, and the F and C lines of hydrogen, respectively. *Synonyms: NU VALUE; VEE VALUE.*

aberration 1. In an optical system, any systematic departure from an idealized path of light rays forming an image, causing the image to be imperfect. 2. In physical optics, any systematic departure of a wave front from its ideal plane or spherical form.
Note: Common aberrations include spherical and chromatic aberration, coma, distortion of image, curvature of field, and astigmatism. *See also: CHROMATIC ABERRATION.*

absolute delay The amount of time by which a signal is delayed. It may be expressed in time (milliseconds, etc.) or in number of characters (pulse times, word times, major cycles, minor cycles, etc.). *See also: DELAY EQUALIZER; DELAY LINE; PHASE DELAY.*

absolute luminance threshold The lowest limit of luminance necessary for vision.

absolute luminosity curve The plot of spectral luminous efficiency versus wavelength.

absolute magnification The magnification produced by a lens placed in front of a normal eye at such a distance from the eye that either the rear focal point of the lens coincides with the center of rotation of the eye or else the front focal point of the eye coincides with the second principal point of the lens. All under the condition that the object is located close to the front focal point of the lens.
Note: This magnification is numerically equal to the distance of distinct vision divided by the equivalent focal length of the lens, with both distances expressed in the same units.

absolute refractive index *See: REFRACTIVE INDEX.*

absolute temperature scale *See: KELVIN TEMPERATURE SCALE.*

absorptance *See: INTERNAL ABSORPTANCE: SPECTRAL ABSORPTANCE.*

absorption In an optical waveguide, that portion of attenuation resulting from conversion of optical power into heat.
Note: Intrinsic components consist of tails of the ultraviolet and infrared absorption bands. Extrinsic components include (a) impurities, e.g., the OH^- ion and transition metal ions and, (b) defects, e.g., results of thermal history and exposure to nuclear radiation. *See also: ATTENUATION.*

absorption coefficient The coefficient in the exponent of the absorption equation that expresses Bouger's Law viz.

$$F = F_0 \, e^{-BX}$$

where F is the electromagnetic (light) flux or intensity at the point X. F_0 is the initial value of flux at $X = 0$, and B is the absorption coefficient.
Note: If an infinitesimally thin layer of absorptive material is considered, making X nearly zero, the absorption coefficient is proportional to the rate of change of flux intensity with respect to distance, i.e., it is proportional to the slope of the absorption curve at that point. The absorption coefficient is a function of wavelength. *See also: ABSORPTIVITY.*

absorption index 1. The ratio of the electromagnetic radiation absorption constant to the refractive index as given by the relation:

$$K' = K\lambda/4\pi n$$

where K is the coefficient of absorption, λ is the wavelength in vacuum, and n is the index of refraction of the absorptive material. 2. The functional relationship between the sun's angle (at any latitude and local time) and the ionospheric absorption.

absorption peak In lightwave transmission media, such as glass, quartz, silica, or plastic, used in optical fibers, slab dielectric waveguides, integrated optical circuits, or other light conducting media, the specific wavelength at which a particular impurity, such as Cu, Fe, Ni, V, Cr, and Mn ions, absorbs the most power, i.e., creates maximum attenuation of the propagated light waves.
Note: Absorption by these impurities at other wavelengths is less than the absorption peak.

absorptive modulation Modulation of a light wave in a medium caused by variation of an applied electric field that causes variations in optical absorption near the edges of the absorption band of the material.

absorptivity The internal absorptance per unit thickness of a medium.
Note: Numerically, absorptivity is unity minus the transmissivity. *See also: ABSORPTION COEFFICIENT.*

acceptance angle Half the vertex angle of that cone within which optical power may be coupled into bound modes of an optical waveguide.
Note: 1. Acceptance angle is a function of position on the entrance face of the core when the refractive index is a function of radius in the core. In that case, the local acceptance angle is

$$\arcsin \sqrt{n^2(r) - n_2^2}$$

where n(r) is the local refractive index and n_2 is the minimum refractive index of the cladding. The sine of the local acceptance angle is sometimes referred to as the local numerical aperture. 2. Power may be coupled into leaky modes at angles exceeding the acceptance angle. *See also: LAUNCH NUMERICAL APERTURE; POWER-LAW INDEX PROFILE.*

acceptance cone A cone whose included apex angle is equal to twice the acceptance angle.

acceptance pattern For an optical fiber or bundle, a curve of total transmitted power plotted against the launch angle.
Note: The total transmitted power or radiation intensity is dependent upon the incident intensity, launch angle (input or incident angle), the transmission coefficient at the fiber interface, and the illumination area.

access *See: CODE DIVISION MULTIPLE ACCESS; DIRECT ACCESS; DUAL ACCESS; DUAL-USE ACCESS LINE; FREQUENCY DIVISION MULTIPLE ACCESS; MAXIMUM ACCESS TIME; MULTIPLE ACCESS; PULSE ADDRESS MULTIPLE ACCESS; REMOTE ACCESS; SERIAL ACCESS; SPECIAL GRADE ACCESS LINE; TIME DIVISION MULTIPLE ACCESS.*

access coupler A device placed between two waveguide ends to allow signals to be withdrawn from or entered into one of the waveguides. *See also: OPTICAL WAVEGUIDE COUPLER.*

accommodation A function of the human eye, whereby its total refracting power, accomplished by a neuromuscular feedback system from the fovea of the retina to muscles that cause the lens to thin or thicken, is varied in order to clearly see objects at different distances.

accommodation limit The distance of the nearest and farthest points, at which an object can be clearly focussed on the retina by the eyes of an observer, usually varying from 4 to 5 inches to infinity.

achromat A compound lens corrected to have the same focal length for two or more wavelengths of light.

achromatic Free from color or hue, such as an optical system free from chromatic aberration.
Note: An achromat is a compound lens corrected to have the same focal length for two or more wavelengths.

achromatic lens A lens, consisting of two or more elements, usually made of crown and flint glass, that has been corrected, so that light of at least two selected wavelengths is focused at a single axial point.

acoustic coupler A device for coupling electrical signals, by acoustical means, usually into and out of a telephone instrument.

acoustic noise An undesired disturbance in the audio frequency range.

acousto-optic effect A periodic variation of refractive index caused by an acoustic wave.

Note: The acousto-optic effect is used in devices that modulate and deflect light. *See also: MODULATION.*

acquisition In satellite communications, the process of locking tracking equipment on a signal from a communications satellite.

acquisition time 1. In a communication system, the amount of time required to attain synchronism. 2. In satellite control communications, the time required for locking tracking equipment on a signal from a communications satellite.

actinometry The science of measurement of radiant energy, particularly that of the sun.

activated chemical vapor deposition process (PACVO) *See: PLASMA-ACTIVATED CHEMICAL VAPOR DEPOSITION PROCESS (PACVD).*

activation energy Coefficient used to characterize the temperature dependence of processes such as degradation of lasers and LEDs, and impurity diffusion. The functional dependence is an exponential of coefficient ($-E_a/kT$), where E_a is the activation energy expressed in electron volts, k is the Boltzmann constant, and T is the absolute temperature expressed in kelvins.

Note: For examples see M. Ettenberg and H. Kressel, *IEEE J. Quantum Electron.,* vol. QE-16, pp. 186–196, 1980. Symbol: E_a.

active laser medium The material within a laser, such as crystal, gas, glass, liquid, or semiconductor, that emits coherent radiation (or exhibits gain) as the result of stimulated electronic or molecular transitions to lower energy states. *Synonym: LASER MEDIUM. See also: LASER; OPTICAL CAVITY.*

active layer The active medium of a semiconductor injection laser or LED. The only part in the device dielectric waveguide that provides optical gain.

active material *See: ACTIVE LASER MEDIUM.*

active network A network that includes a source of power. *See also: PASSIVE NETWORK.*

active optics Pertaining to the development and use of optical components whose characteristics are controlled during their operational use in order to modify characteristics, such as wave front direction, polarization, mode, intensity, or path, of an electromagnetic wave in the visible or near visible region of the frequency spectrum; in contrast to inactive, rigid, or fixed optics in which components are not varied, with primary attention being given to measurement and control of wavefronts or rays in real time in order to concentrate radiated energy on a detector, target, waveguide, or other device.

active region *See: ACTIVE LAYER.*

active satellite A satellite that transmits a signal, in contrast to a passive satellite that only reflects a signal.

ACVD *Acronym for ACTIVATED CHEMICAL VAPOR DEPOSITION PROCESS.*

adaptation *See: DARK ADAPTATION; LIGHT ADAPTATION.*

adaptive channel allocation A method of multiplexing wherein the information handling capacities of channels are not predetermined but are assigned on demand. *See also: CHANNEL; MULTIPLEX.*

adaptive predictive coding (APC) A narrowband analog-to-digital conversion technique employing a one-level or multilevel sampling system in which the value of the signal at each sample time is adaptively predicted to be a linear function of the past values of the quantized signals.

Note: APC is related to linear predictive coding (LPC) in that both use adaptive predictors. However, APC uses fewer prediction coefficients, thus requiring a higher information bit rate than LPC. *See also: LINEAR PREDICTIVE CODING.*

adaptive technique *See: COHERENT OPTICAL ADAPTIVE TECHNIQUE.*

additive white Gaussian noise (AWGN) *Synonym: WHITE NOISE.*

adjacent-channel interference Extraneous power from an authorized signal in an adjacent channel. *See also: CHANNEL.*

APD coupler *See: AVALANCHE PHOTODIODE COUPLER.*

air-spaced doublet In optics, a compound lens of two elements with air or empty space between them.

AlGaAs *Chemical symbol for ALUMINUM GALLIUM ARSENIDE.*

AlGaAsSb *Chemical symbol for ALUMINUM GALLIUM ARSENIDE ANTIMONIDE.*

AlGaSb *Chemical symbol for ALUMINUM GALLIUM ANTIMONIDE.*

aligned bundle A bundle of optical fibers in which the relative spatial coordinates of each fiber are the same at the two ends of the bundle.

Note: The term "coherent bundle" is often employed as a synonym, and should not be confused with phase coherence or spatial coherence. *Synonym: COHERENT BUNDLE. See also: FIBER BUNDLE.*

alpha profile *See: POWER-LAW INDEX PROFILE.*

alternate mark inversion signal (AMI signal) A pseudo-ternary signal, conveying binary digits, in which successive "marks" are of alternative polarity (positive and negative) but equal in amplitude and in which "spaces" are of zero amplitude. *Synonym: BIPOLAR SIGNAL. See also: AMI VIOLATION; MODIFIED AMI; PAIRED-DISPARITY CODE.*

alternating code *Synonym: PAIRED/DISPARITY CODE.*

aluminum gallium antimonide A crystalline semiconductor ternary alloy between aluminum antimonide and gallium antimonide. Lattice-matched to GaSb substrates over the wavelength range: 1.4–1.8 μm. Used mostly for long-wavelength APDs.

Note: See for example H. D. Law, *et al., IEEE J. Quantum Electron.,* vol. QE-15, pp. 549–559, 1979.

aluminum gallium arsenide A crystalline semiconductor alloy between gallium arsenide and aluminum arsenide, which is grown in layer form atop GaAs substrates to provide optical and carrier confinement in double-heterostructure and single-heterostructure diode lasers for the 0.7–0.9 μm wavelength region. *Symbols:* AlGaAs; $Al_xGa_{1-x}As$; (Al, Ga)As.

aluminum gallium arsenide antimonide A crystalline semiconductor quaternary alloy between aluminum arsenide, gallium arsenide, aluminum antimonide and gallium antimonide. Written as $Al_xGa_{1-x}As_ySb_{1-y}$. Grown lattice-matched to GaSb substrates to provide the light-confinement layers and/or active layers for lasers and LEDs emitting in the 1.2 to 1.8 μm range. Also used for long-wavelength APD and PIN detectors. *Symbols:* AlGaAsSb; (Al,Ga)(As,Sb), $Al_xGa_{1-x}As_ySb_{1-y}$.

aluminum garnet source *See: YAG/LED SOURCE.*

AM *Acronym for AMPLITUDE MODULATION.*

ambient noise level The level of acoustic noise existing in a room or other location, as measured with a sound level meter. It is usually measured in decibels above a reference level of 0.00002 newton per square meter in SI units, or 0.0002 dyne per square centimeter in cgs units. *Synonym: ROOM NOISE LEVEL. See also: NOISE.*

ambient temperature The temperature of air or other media surrounding equipment.

AME *Acronym for AMPLITUDE MODULATION EQUIVALENT.*

AMI *Acronym for ALTERNATE MARK INVERSION.*

AMI violation A "mark" which has the same polarity as the previous "mark" in the transmission of alternate mark inversion (AMI) signals. *Synonym: BIPOLAR VIOLATION. See also: ALTERNATE MARK INVERSION SIGNAL.*

amplification by stimulated emission of radiation *See: MICROWAVE AMPLIFICATION BY STIMULATED EMISSION OF RADIATION.*

amplifier noise The thermal noise figure of preamplifiers used in conjunction with photodetectors. The main limiting factor of the receiver sensitivity when using PIN detectors and/or in the case of low signal-to-noise, high-frequency applications.

amplitude distortion Distortion occurring in an amplifier or other device when the output amplitude is not a linear function of the input amplitude under specified conditions.

Note: Amplitude distortion is measured with the system operating under steady-state conditions with a sinusoidal input signal. When other frequencies are present, the term amplitude refers to that of the fundamental only.

amplitude equalizer A corrective network which is designed to modify the amplitude characteristics of a circuit or system over a desired frequency range.

amplitude frequency response *Synonym: INSERTION LOSS VS FREQUENCY CHARACTERISTIC.*

amplitude hit In a data transmission channel, a momentary disturbance caused by a sudden change in the amplitude of the signal. *See also: HIT.*

amplitude modulation (AM) A form of modulation in which the amplitude of a carrier wave is varied in accordance with some characteristic of the modulating signal.

amplitude modulation equivalent (AME) *Synonym: COMPATIBLE SIDE-BAND TRANSMISSION.*

amplitude quantized control *Synonym: AMPLITUDE QUANTIZED SYNCHRONIZATION.*

amplitude quantized synchronization A synchronization control system in which the functional relationship between actual phase error and derived error signal includes discontinuities. *Note:* In practice this implies that the working range of phase errors is divided into a finite number of subranges and that a unique signal is derived for each subrange whenever the error falls within a subrange. *Synonym: AMPLITUDE QUANTIZED CONTROL.*

amplitude-versus-frequency distortion That distortion in a transmission system caused by the nonuniform attenuation, or gain, of the system with respect to frequency under specified terminal conditions. *Synonym: FREQUENCY DISTORTION.*

analog control *Synonym: ANALOG SYNCHRONIZATION.*

analog data Data represented by a physical quantity that is considered to be continuously variable and whose magnitude is made directly proportional to the data or to a suitable function of the data.

analog decoding A process in which one of a set of reconstructed analog signal samples is generated from the digital signal representing a sample. *See also: ANALOG ENCODING.*

analog encoding A process in which digital signals are generated, representing the sample taken of an analog signal value at a given instant. *See also: ANALOG DECODING; UNIFORM ENCODING.*

analog-intensity modulation In an optical modulator, the variation of the intensity, i.e. instantaneous output power level, of a light source in accordance with an intelligence-bearing signal or continuous wave, the resulting envelop normally being detectable at the other end of a light-wave transmission system.

analog signal A nominally continuous electrical signal that varies in some direct correlation to a nonelectrical signal impressed on a transducer. The electrical signal may vary its frequency or amplitude, for instance, in response to change in phenomena or characteristics such as sound, light, heat, position, or pressure.

analog switch A switching equipment designed, designated, or used to connect circuits between users for real-time transmission of analog signals.

analog synchronization A synchronization control system in which the relationship between the actual phase error between clocks and the error signal device is a continuous function over a given range. *Synonym: ANALOG CONTROL. See also: SYNCHRONIZATION.*

analog-to-digital (A-D) coder *Synonym: ANALOG-TO-DIGITAL ENCODER.*

analog-to-digital (A-D) converter A device which converts an analog input signal to a digital output signal carrying equivalent information. *See also: DIGITAL-TO-ANALOG (D-A) CONVERTER.*

analog-to-digital (A-D) encoder A device for encoding analog signal samples. *Synonyms: ANALOG-TO-DIGITAL CODER; CODER.*

analyzer *See: LIGHT ANALYZER.*

anamorphic In optical systems, pertaining to a configuration of optical components, such as lenses, mirrors, and prisms, that produce different effects on an image in different directions or different effects on different parts, for example producing different magnification in different directions or converting a point on an object to a line on its image.

angle *See: ACCEPTANCE ANGLE; BREWSTER ANGLE; CONVERGENCE ANGLE; CRITICAL ANGLE; DEVIATION ANGLE; EXIT ANGLE; LAUNCH ANGLE; LIMITING RESOLUTION ANGLE; MAXIMUM ACCEPTANCE ANGLE; REFLECTION ANGLE; REFRACTION ANGLE.*

angle - between - half - power - points *See: EMISSION BEAM ANGLE BETWEEN HALF-POWER POINTS.*

angle modulation Modulation in which the angle of a sine wave carrier is varied. Phase and frequency modulation are particular forms of angle modulation.

angle of deviation In optics, the net angular deflection experienced by a light ray after one or more refractions or reflections. *Note:* The term is generally used in reference to prisms, assuming air interfaces. The angle of deviation is then the angle between the incident ray and the emergent ray. *See also: REFLECTION; REFRACTION.*

angle of incidence The angle between an incident ray and the normal to a reflecting or refracting surface. *See also:*

CRITICAL ANGLE; TOTAL INTERNAL REFLECTION.

angstrom (Å) A unit of optical wavelength (obsolete). 1 Å = 10^{-10} meters.
Note: The angstrom has been used historically in the field of optics, but it is not an SI (International System) unit.

angular magnification The ratio of the apparent size of an image seen through an optical element or instrument to that of the object viewed by the unaided eye, when both the object and image are at infinity, which is the case for telescopes, or when both the object and image are considered to be at the distance of distinct vision, which is the case for microscopes. *Synonym:* MAGNIFYING POWER.

angular misalignment loss The optical power loss caused by angular deviation from the optimum alignment of source to optical waveguide, waveguide to waveguide, or waveguide to detector. *See also: EXTRINSIC JOINT LOSS; GAP LOSS; INTRINSIC JOINT LOSS; LATERAL OFFSET LOSS.*

anisochronous Pertaining to transmission in which the time interval separating any two significant instants in sequential signals is not necessarily related to the time interval separating any other two significant instants. *See also: HETEROCHRONOUS, HOMOCHRONOUS, ISOCHRONOUS, MESOCHRONOUS, PLESIOCHRONOUS.*

anisochronous transmission *See: ASYNCHRONOUS TRANSMISSION.*

anisotropic Pertaining to a material whose electrical or optical properties are different for different directions of propagation or different polarizations of a traveling wave. *See also: ISOTROPIC.*

anomalous propagation (AP) Abnormal propagation due to discontinuities in the propagation medium and resulting, in many instances, in the reception of signals well beyond their normal range.

answering The process of responding to a call to complete a connection. *See also: AUTOMATIC ANSWERING.*

antenna Any structure or device used to collect or radiate electromagnetic waves. *See also: APERIODIC ANTENNA; APERTURE; APERTURE-TO-MEDIUM COUPLING LOSS; BEAM-WIDTH; BILLBOARD ANTENNA; DIPOLE ANTENNA; DIRECTIVE GAIN; EFFECTIVE ANTENNA LENGTH; EFFECTIVE HEIGHT;* EFFECTIVE RADIATED POWER; IMAGE ANTENNA; ISOTROPIC ANTENNA; LIGHT ANTENNA; LOG-PERIODIC ANTENNA; OMNI-DIRECTIONAL ANTENNA; PARABOLIC ANTENNA; PERISCOPE ANTENNA; PHASED ARRAY; POWER GAIN OF AN ANTENNA; RHOMBIC ANTENNA; SLOT ANTENNA; TEST ANTENNA.

antenna gain *Synonym:* POWER GAIN OF AN ANTENNA.

antenna gain-to-noise temperature (G/T) For satellite earth terminal receiving systems, a figure of merit which equals G/T, where G is the gain in dB of the earth terminal antenna at the receive frequency, and T is the equivalent noise temperature of the receiving system in kelvins. *See also: GAIN; NOISE.*

antenna height above average terrain The average of antenna heights above the terrain from two to ten miles from the antenna for the eight directions spaced evenly for each 45° of azimuth starting with true north.
Note: In general, a different antenna height will be determined in each direction from the antenna. The average of these various heights is considered the antenna height above average terrain. (In some cases, fewer than eight directions may be used.)

antenna lobe A three-dimensional section of the radiation pattern of a directional antenna bounded by one or two cones of nulls or regions of diminished intensity.

antenna matching The process of adjusting impedance so that the input impedance of an antenna equals the characteristic impedance of its transmission line.

antenna noise temperature The temperature of a resistor having an available noise power per unit bandwidth equal to that at the antenna output at a specified frequency.
Note: Noise temperature of an antenna depends on its coupling to all noise sources in its environment as well as noise generated within the antenna. *See also: NOISE.*

antiguide Dielectric waveguide in which light is partially lost by refraction upon grazing incidence at the dielectric boundaries. Also called "leaky" guide. In active structures such as lasers antiguiding losses can be compensated by electronic gain. Important in lasing structures as it provides strong discrimination against high-order mode oscillation. *Synonym: LEAKY WAVEGUIDE.*

antinode A point in a standing or stationary wave at which the amplitude is a maximum.
Note: The wave should be identified as a voltage wave or a current wave. *See also: NODE; STANDING WAVE RATIO (SWR).*

antireflection coating A thin, dielectric or metallic film (or several such films) applied to an optical surface to reduce the reflectance and thereby increase the transmittance.
Note: The ideal value of the refractive index of a single layered film is the square root of the product of the refractive indices on either side of the film, the ideal optical thickness being one quarter of a wavelength. *See also: DICHROIC FILTER; FRESNEL REFLECTION; REFLECTANCE; TRANSMITTANCE.*

AP *Acronym for ANOMALOUS PROPAGATION.*

APC *Acronym for ADAPTIVE PREDICTIVE CODING.*

APD *Acronym for AVALANCHE PHOTODIODE.*
Note: apd and a.p.d. are also used.

aperiodic antenna An antenna designed to have an approximately constant input impedance over a wide range of frequencies, e.g., terminated rhombic antennas and wave antennas. *Synonym: NONRESONANT ANTENNA.*

aperture 1. In an optical system an opening or hole, through which light or matter may pass that is equal to the diameter of the largest entering beam of light that can travel completely through the system and which may or may not be equal to the aperture of the objective. 2. That portion of a plane surface near a unidirectional antenna, perpendicular to the direction of maximum radiation intensity, through which the major part of the radiation passes. *See also: NUMERICAL APERTURE.*

aperture ratio The aperture ratio R_A is defined as

$$R_A = 2N \sin A$$

where N is the refractive index of the image space, and A is the maximum angular opening of the axial bundle of refracted rays.

Note: The speed, i.e., energy per unit area of images of an objective is related to its aperture ratio as follows:

$$\text{Speed} \sim R_A^2.$$

When the angular opening is small, $N=1$, and object distance is great, it is approximately true that

$$N \sin A = \frac{D}{2F}$$

or that

$$\frac{F}{D} = F-\text{Number} = \frac{1}{2R_A}$$
$$= \frac{1}{\text{Aperture Ratio}} .$$

aperture stop The physical diameter that limits the size of the cone of radiation that an optical system will accept from an axial point on an object.

aperture-to-medium coupling loss The difference between the theoretical gain of a very large antenna (as used in beyond-the-horizon microwave links) and the gain that can be realized in operation. It is related to the ratio of the scatter angle to the antenna beamwidth.
Note: The "very large antennas" are referred to in wavelengths; thus this loss can apply to LOS systems also. *See also: ANTENNA.*

aplanatic lens A lens that has been corrected for spherical aberration, departure from the sine condition, coma, and color.

array A set of equally spaced sources or detectors placed linearly. The array units can be phase-coupled or independent.

artificial pupil A diaphragm, or other limitation, that confines a beam of light to a smaller cone, for example a pupil that confines a beam of light entering the eye to a smaller cone than does the iris of the human eye.

ASCII (American National Standard Code for Information Interchange) The standard code, using a coded character set consisting of 7-bit coded characters (8 bits including parity check), used for information interchange among data processing systems, data communication systems, and associated equipment.
Note: The ASCII set consists of control characters and graphic characters, and is properly an alphabet not a code. It is the U.S. implementation of ITA No. 5. *See also: ALPHABET; CODE.*

aspect *See: IMAGE ASPECT.*

assembly *See:* CABLE ASSEMBLY; MUL-TIPLE-BUNDLE CABLE ASSEMBLY; MULTIPLE-FIBER CABLE ASSEMBLY; OPTICAL HARNESS ASSEMBLY.

assigned frequency The frequency of the center of the radiated bandwidth. *Note:* The frequency of the RF carrier, whether suppressed or radiated, is usually given in parentheses following the assigned frequency, and is the frequency appearing in the dial settings of RF equipment intended for single sideband or independent sideband transmission. *See also:* AUTHORIZED FREQUENCY; FREQUENCY; FREQUENCY ALLOCATION; FREQUENCY ASSIGNMENT.

astigmatism An aberration of a lens or lens system that causes an off-axis point to be imaged as two separated lines perpendicular to each other.

asymmetrical modulator *Synonym:* UNBALANCED MODULATOR.

asynchronous communication system A data communication system in which extra signal elements are appended to the data for the purpose of synchronizing individual data characters or blocks. *Note:* The time spacing between successive data characters or blocks may be of arbitrary duration. *Synonym:* START-STOP SYSTEM.

asynchronous network *Synonym:* NON-SYNCHRONOUS NETWORK.

asynchronous operation 1. A sequence of operations in which operations are executed out of time coincidence with any event. 2. An operation that occurs without a regular or predictable time relationship to a specified event, e.g., the calling of an error diagnostic routine that may receive control at any time during the execution of a computer program. *Synonym:* ASYNCHRONOUS WORKING.

asynchronous time-division multiplexing (ATDM) An asynchronous transmission mode that makes use of time-division multiplexing. *See also:* SYNCHRONOUS TDM; TIME-DIVISION MULTIPLEX.

asynchronous transmission A transmission process such that between any two significant instants in the same group (in data transmission this group is a block or a character; in telegraphy this group is a character) there is always an integral number of unit intervals. Between two significant instants located in different groups there is not always an integral number of unit intervals. *See also:* ISOCHRONOUS; PLESIOCHRONOUS; SYNCHRONOUS TRANSMISSION.

asynchronous working *Synonym:* ASYNCHRONOUS OPERATION.

atmosphere laser *See:* LONGITUDINALLY EXCITED ATMOSPHERE LASER; TRANSVERSE EXCITED ATMOSPHERE LASER.

atmospheric duct A layer in the lower atmosphere, occasionally of great horizontal extent, in which the vertical refractivity gradients are such that radio signals are guided or focused within the duct and tend to follow the curvature of the earth with much less than normal attenuation.

atmospheric noise Radio noise caused by natural atmospheric processes, primarily by lightning discharges in thunderstorms. *See also:* NOISE.

atomic defect absorption In lightwave transmission media such as optical fibers and integrated optical circuits made of glass, silica, plastic, and other materials, the absorption of light energy from a traveling or standing wave by atomic changes, brought about during or after production, by exposure to radiation, such as, infrared or gamma radiation at high levels. For example titanium-doped silica can develop losses of several thousand dB/kilometer when the fibers are drawn under high temperature, and conventional fiber optic glasses can develop losses of 20,000 dB/kilometer during and after exposure to gamma radiation of 3000 rads.

attack time In an echo suppressor, the time interval between the instant that the signal level at the receive "in" port exceeds the suppression activate point and the instant when the suppression loss is introduced.

attenuation In an optical waveguide, the diminution of average optical power. *Note:* In optical waveguides, attenuation results from absorption, scattering, and other radiation. Attenuation is generally expressed in dB. However, attenuation is often used as a synonym for attenuation coefficient, expressed in dB/km. This assumes the attenuation coefficient is invariant with length. *See also:* ATTENUATION COEFFICIENT; COUPLING LOSS; DIFFERENTIAL MODE ATTENUATION; EQUILIBRIUM MODE DISTRIBUTION; EXTRINSIC JOINT LOSS; IN-

SERTION LOSS; INTRINSIC JOINT LOSS; LEAKY MODES; MACROBEND LOSS; MATERIAL SCATTERING; MICRO- BEND LOSS; RAYLEIGH SCATTERING; SPECTRAL WINDOW; TRANSMISSION LOSS; WAVEGUIDE SCATTERING.

attenuation coefficient The rate of diminution of average optical power with respect to distance along the wave- guide. Defined by the equation

$$P(z) = P(0)\, 10^{-(\alpha z/10)}$$

where $P(z)$ is the power at distance z along the guide and $P(0)$ is the power at $z = 0$; α is the attenuation coefficient in dB/km if z is in km. From this equation,

$$\alpha z = -10 \log_{10} [P(z)/P(0)].$$

This assumes that α is independent of z; if otherwise, the definition must be giv- en in terms of incremental attenuation as:

$$P(z) = P(0)10^{-\int_0^z \alpha(z)dz/10}$$

or, equivalently,

$$\alpha(z) = -10\frac{d}{dz} \log_{10} [P(z)/P(0)]$$

See also: ATTENUATION; ATTENUA- TION CONSTANT; AXIAL PROPAGA- TION CONSTANT.

attenuation constant For a particular mode, the real part of the axial propaga- tion constant. The attenuation coeffi- cient for the mode power is twice the attenuation constant. *See also: ATTENU- ATION COEFFICIENT, AXIAL PROPA- GATION CONSTANT; PROPAGATION CONSTANT.*

attenuation-limited operation The condition prevailing when the received signal amplitude (rather than distortion) limits performance. *See also: BAND- WIDTH-LIMITED OPERATION; DIS- TORTION-LIMITED OPERATION.*

attenuation term In the propagation of an electromagnetic wave in a waveguide such as an optical fiber, thin film, or metal pipe, the term A in the expression for the exponential variation characteris- tic of guided waves,

$$e^{-PZ} = e^{-Z(IH+A)}$$

represents the attenuation or pulse am- plitude diminution experienced per unit of propagation distance of the wave.

Note: In a given guide, the phase term H is initially assumed to be independent of the attenuation term, A, which is then found separately assuming H does not change with losses. For an optical fiber, the attenuation term A is supplied by the manufacturer since it can be experi- mentally measured. *See also: PHASE TERM; PROPAGATION CONSTANT.*

attenuator *See: CONTINUOUS VARI- ABLE OPTICAL ATTENUATOR; FIXED OPTICAL ATTENUATOR; OPTICAL ATTENUATOR; STEPWISE VARIABLE OPTICAL ATTENUATOR.*

Auger recombination Nonradiative re- combination process during which the energy released via electron-hole recom- bination is used to excite another elec- tron in the conduction band or another hole in the valence band. Believed to be the main cause of strong threshold-cur- rent temperature sensitivity (i.e., low T_0 values) and sublinear LED light output characteristics for long-wavelength de- vices.

Note: For example see N. K. Dutta and R. J. Nelson, *IEEE J. Quantum Electron.,* vol. QE-18, pp. 871–878, 1982.

auxiliary channel In data transmission, a secondary channel whose direction of transmission is independent of the pri- mary channel and is controlled by an appropriate set of secondary control in- terchange circuits. *See also: CHANNEL.*

avalanche gain Current gain obtained in avalanche photodiodes (APDs) by multiplicative generation of electrons and holes via impact ionization at high electric field values. Symbol: M. Typical M values for maximum signal-to-noise (S/N) ratio are around 10. For larger M values the S/N degrades, and as M ap- proaches infinity avalanche breakdown occurs. M is the mean-square value of the diode's internal gain. Due to ava- lanche gain, APDs are much more sensi- tive than PIN photodiodes. *Synonym: MULTIPLICATION FACTOR.*

avalanche photodiode (APD) A pho- todiode designed to take advantage of avalanche multiplication of photocur- rent.

Note: As the reverse-bias voltage ap- proaches the breakdown voltage, hole– electron pairs created by absorbed pho- tons acquire sufficient energy to create additional hole–electron pairs when they collide with ions; thus a multiplication (signal gain) is achieved. *See also: PHO- TODIODE; PIN PHOTODIODE.*

avalanche photodiode coupler A coupling device that enables the coupling of light energy from an optical fiber onto the photo sensitive surface of an avalanche photodiode (APD) of a photon detector (photodetector) at the receiving end of an optical fiber data link.
Note: The coupler may be only a fiber pigtail epoxied to the APD. *Synonym: APD COUPLER.*

average power In a pulsed laser, the energy per pulse (joules) times the pulse repetition rate (hertz). Usually in watts.

average rate of transmission *Synonym: EFFECTIVE SPEED OF TRANSMISSION.*

AVPO *Acronym for AXIAL VAPOR-PHASE OXIDATION PROCESS.*

AWGN *Acronym ADDITIVE WHITE GAUSSIAN NOISE.*

axial bundle A cone of electromagnetic rays, such as light rays, that emanate from an object point that is located on the optical axis of a lens system.

axial mode *See: LONGITUDINAL MODE.*

axial propagation constant The propagation constant evaluated along the axis of a waveguide (in the direction of transmission).
Note: The real part of the axial propagation constant is the attenuation constant while the imaginary part is the phase constant. *Synonym: AXIAL PROPAGATION WAVE NUMBER. See also: ATTENUATION; ATTENUATION COEFFICIENT; ATTENUATION CONSTANT; PROPAGATION CONSTANT.*

axial propagation wave number *Synonym: AXIAL PROPAGATION CONSTANT.*

axial ray A light ray that travels along the optical axis. *See also: GEOMETRIC OPTICS; FIBER AXIS; MERIDIONAL RAY; PARAXIAL RAY; SKEW RAY.*

axial slab interferometry *Synonym: SLAB INTERFEROMETRY.*

axial vapor-phase oxidation process (AVPO) A vapor-phase oxidation (VPO) process for making Graded-Index (GI) optical fibers in which the glass preform is grown radially rather than longitudinally as in other processes, the refractive index thus being controlled in a spatial domain rather than a time domain, and the chemical gases are burned in an oxyhydrogen flame, as in the OVPO process, to produce a stream of soot particles to produce the graded index of refraction.

axis *See: OPTICAL AXIS.*

axis paraboloidal mirror *See: OFF-AXIS PARABOLOIDAL MIRROR.*

back focal length (BFL) The distance measured from the vertex of the back surface of a lens to its rear focal point.

background noise The total system noise in the absence of information transmission; it is independent of the presence or absence of a signal.

backscatter 1. The deflection by reflection or refraction of an electromagnetic wave or signal in such a manner that a component of the wave is deflected opposite to the direction of propagation of the incident wave or signal. 2. The component of an electromagnetic wave or signal that is deflected by reflection or refraction opposite to the direction of propagation of the incident wave or signal. 3. To deflect, by reflection or refraction, an electromagnetic wave or signal in such a manner that a component of the wave or signal is deflected opposite to the direction of propagation of the incident wave or signal.

Note: The term scatter can be applied to reflection or refraction by relatively uniform media but it is usually taken to mean propagation in which the wavefront and direction are modified in a relatively disorderly fashion. *See also: FORWARD SCATTER; PROPAGATION; SCATTER.*

backscattering The scattering of light into a direction generally reverse to the original one. *See also: RAYLEIGH SCATTERING; REFLECTANCE; REFLECTION.*

back-surface mirror An optical mirror on which the reflecting surface is applied to the back surface of the mirror, i.e. not to the surface of first incidence.

Note: The reflected light must pass through the substrate twice, once as part of the incident light and once as the reflected light. *See also: FRONT-SURFACE MIRROR.*

back-to-back connection A connection between the output of a transmit-ting device and the input of an associated receiving device.

Note: This eliminates the effects of the transmission channel or medium.

backward channel 1. In data transmission, a secondary channel whose direction of transmission is constrained to be opposite to that of the primary (or forward) channel. The direction of transmission in the backward channel is restricted by the control interchange circuit that controls the direction of transmission in the primary channel. 2. That channel of a data circuit which passes data in a direction opposite to that of its associated forward channel.

Note: The backward channel is usually used for transmission of supervisory, acknowledgement, or error-control signals. The direction of flow of these signals is in the direction opposite to that in which information is being transferred. The bandwidth of this channel is usually less than that of the forward channel, i.e., the information channel. *See also: DATA TRANSMISSION; FORWARD CHANNEL.*

backward signal A signal sent in the direction from the called to the calling station, or from a data sink to a data source.

Note: The backward signal is usually sent in the backward channel.

backward supervision The use of supervisory sequences from a secondary to a primary station. *See also: SECONDARY STATION.*

balanced Pertaining to electrical symmetry.

balanced code In PCM, a code whose digital sum variation is finite.

Note: Balanced codes have no dc component in their frequency spectrum. *See also: CODE; PULSE-CODE MODULATION.*

balanced line A transmission line consisting of two conductors in the presence of ground capable of being operated in such a way that when the voltages of the two conductors at all transverse planes are equal in magnitude and opposite in polarity with respect to ground, the currents in the two conductors are equal in magnitude and opposite in direction. *Synonym: BALANCED SIGNAL PAIR.*

balanced signal pair *Synonym: BALANCED LINE.*

balanced station That station responsible for performing balanced link-level operation.
Note: A balanced station generates commands and interprets responses, and interprets received commands and generates responses.

balancing network 1. A circuit used to simulate the impedance of a uniform two-wire cable or open-wire circuit over a selected range of frequencies. 2. A device used between a balanced device or line and an unbalanced device or line for the purpose of transforming from balanced to unbalanced or from unbalanced to balanced.

balsam *See: CANADA BALSAM.*

balun An RF balancing network for coupling between balanced and unbalanced devices, normally used between equipment and transmission lines or transmission lines and antennas.

band 1. In communication, the frequency spectrum between two defined limits. 2. A group of tracks on a magnetic drum or on one side of a magnetic disc. *See also: FREQUENCY; CONDUCTION BAND; FREQUENCY GUARD BAND; INFRARED BAND; TIME GUARD BAND; VALENCE BAND.*

band-elimination filter *Synonym: BANDSTOP FILTER.*

bandgap energy The size of the energy gap across which light is generated or absorbed in semiconductor devices. Typically quoted in electron volts.

bandgap step The difference in bandgap energy between the confining layer(s) and the active layer of a semiconductor laser. A large bandgap step ensures tight injected-carrier confinement and consequently low threshold current densities.

bandpass filter *See: OPTICAL FILTER.*

band-rejection filter *Synonym: BANDSTOP FILTER.*

band-stop filter A filter having a single continuous attenuation band, neither the upper nor lower cutoff frequencies being either zero or infinite. *Synonyms: BAND-ELIMINATION FILTER; BAND-REJECTION FILTER; BAND-SUPPRESSION FILTER.*

band-suppression filter *Synonym: BAND-STOP FILTER.*

bandwidth *See: FIBER BANDWIDTH.*

bandwidth-limited operation The condition prevailing when the system bandwidth, rather than the amplitude (or power) of the signal, limits performance. The condition is reached when the system distorts the shape of the waveform beyond specified limits. For linear systems, bandwidth-limited operation is equivalent to distortion-limited operation. *See also: ATTENUATION-LIMITED OPERATION; DISTORTION-LIMITED OPERATION; LINEAR OPTICAL ELEMENT.*

bandwidth product *See: GAIN–BANDWIDTH PRODUCT.*

barrier layer In the fabrication of an optical fiber, a layer that can be used to create a boundary against OH^- ion diffusion into the core. *See also: CORE.*

barrier-layer cell *See: PHOTOVOLTAIC CELL.*

baseband In the process of modulation, the frequency band occupied by the aggregate of the transmitted signals when first used to modulate a carrier, and in demodulation, the recovered aggregate of the transmitted signals.
Note: The term is commonly applied to cases where the ratio of the upper to the lower limit of the frequency band is large compared to unity. *See also: MULTIPLEX BASEBAND; MULTIPLEX BASEBAND RECEIVE TERMINALS; MULTIPLEX BASEBAND SEND TERMINALS; RADIO BASEBAND; RADIO BASEBAND RECEIVE TERMINALS; RADIO BASEBAND SEND TERMINALS.*

baseband response function *Synonym: TRANSFER FUNCTION (of a device).*

basic group *See: GROUP.*

basic mode *See: LASER BASIC MODE.*

basic status In data transmission, a secondary station's capability to send or receive a frame containing an information field. *See also: DATA TRANSMISSION.*

baud (Bd) 1. A unit of modulation rate. One baud corresponds to a rate of one unit interval per second, where the modulation rate is expressed as the reciprocal of the duration in seconds of the unit interval. *Example:* If the duration of the unit interval is 20 milliseconds, the modulation rate is 50 baud. 2. A unit of signaling speed equal to the number of discrete signal conditions, variations, or events per second.

Baudot code Historically, a five-unit synchronous code which has been replaced by the start-stop asynchronous International Telegraph Alphabet No. 2 (ITA #2). The term Baudot code should not be used to identify ITA #2.

BCI *Acronym for BIT COUNT INTEGRITY.*

beam A shaft or column of electromagnetic radiation, such as radio waves or light or a bundle of rays, consisting of parallel, converging, or diverging rays. *See also: DIVERGING BEAM.*

beam-angle-between-half-power-points *See: EMISSION BEAM-ANGLE-BETWEEN-HALF-POWER-POINTS.*

beam diameter The distance between two diametrically opposed points at which the irradiance is a specified fraction of the beam's peak irradiance; most commonly applied to beams that are circular or nearly circular in cross section. *Synonym: BEAMWIDTH. See also: BEAM DIVERGENCE.*

beam divergence 1. For beams that are circular or nearly circular in cross section, the angle subtended by the far-field beam diameter. 2. For beams that are not circular or nearly circular in cross section, the far-field angle subtended by two diametrically opposed points in a plane perpendicular to the optical axis, at which points the irradiance is a specified fraction of the beam's peak irradiance. Generally, only the maximum and minimum divergences (corresponding to the major and minor diameters of the far-field irradiance) need be specified. *See also: BEAM DIAMETER; COLLIMATION; FAR-FIELD REGION.*

beam divergence angle Half the vertex angle of that cone that encompasses a circle of diameter equal to the beam diameter at all points in the far field. Beam divergence angle is strictly meaningful only when describing the far field of beams that are circular or nearly circular in cross section. *See also: BEAM DIAMETER; BEAM SPREAD; BEAMWIDTH; COLLIMATION; FAR-FIELD REGION.*

beamsplitter A device for dividing an optical beam into two or more separate beams; often a partially reflecting mirror.

beam spread The angle between two planes, within which passes a specified fraction of the total power in an optical beam. Beam spread is generally specified for two orthogonal orientations.
Note: This definition is useful in characterizing beams that are not circular in cross section. *See also: BEAM DIVERGENCE ANGLE.*

beamwidth *Synonym: BEAM DIAMETER.*

Beer's law In the transmission of electromagnetic radiation through a liquid solution (non-absorbing non-scattering solvent containing an absorbing or scattering solute), the attenuation, reduction, decay, or diminution of electromagnetic field intensity or power density is an exponential decay function of the product of the concentration of the solute C, the spectral absorption/scattering coefficient per unit of concentration per unit of distance A, and the thickness X, given by the relationship

$$I = I_0 \, e^{-CAX}$$

where I is the power density at distance X and I_0 is the power density at X=0. *See also: BOUGER'S LAW; LAMBERT'S LAW.*

BER *Acronym for BIT ERROR RATE.*

BFL *Acronym for BACK FOCAL LENGTH.*

BH laser *See: BURIED HETEROSTRUCTURE DIODE LASER.*

BH-LOC laser *See: BURIED HETEROSTRUCTURE LARGE-OPTICAL-CAVITY DIODE LASER.*

bias 1. A systematic deviation of a value from a reference value. 2. The amount by which the average of a set of values departs from a reference value. 3. Electrical, mechanical, or magnetic force that is applied to a relay, vacuum tube, or other device, to establish an electrical or mechanical reference level to operate the device. 4. Effect on telegraph signals produced by the electrical characteristics of the terminal equipment. *See also: MARKING BIAS; SPACING BIAS.*

bias distortion Distortion affecting a two-condition (or binary) modulation in which all the significant intervals corresponding to one of the two significant conditions have uniformly longer or shorter durations than the corresponding theoretical durations. *See also: DISTORTION; INTERNAL BIAS.*

bidirectional transmission Signal transmission in both directions along an optical waveguide or other component.

bifocal In optics, pertaining to a system or component, such as a lens or lens system, that has, or is characterized by, two or more optical foci.

bilateral control *Synonym: BILATERAL SYNCHRONIZATION.*

bilateral synchronization A synchronization control system between exchanges A and B in which the clock at exchange A controls that at exchange B, and the clock at exchange B controls that at exchange A. *Synonym: BILATERAL CONTROL. See also: SYNCHRONIZATION.*

billboard antenna A broadside antenna array with flat reflectors. *See also: ANTENNA.*

bimolecular recombination Band-to-band recombination process for which the injected carrier density substantially exceeds the background concentration. Of importance at high injection levels in all carrier-confining structures of the double-heterojunction type. When dominant, bimolecular recombination determines the excess carrier lifetime. The main reason for the high frequency response of AlGaAs edge-emitting LEDS.

binary code A code composed by selection and configuration of an entity which can assume either one of two possible states.

binary digit (bit) 1. In pure binary notation, either of the characters, 0 or 1. 2. A unit of information equal to one binary decision or the designation of one of two possible and equally likely states of anything used to store or convey information.

binary modulation The process of varying a parameter of a carrier as a function of two finite and discrete states.

bipolar signal *Synonym: ALTERNATE MARK INVERSION SIGNAL.*

bipolar violation *Synonym: AMI VIOLATION.*

birefringence *See: BIREFRINGENT MEDIUM.*

birefringent medium A material that exhibits different indices of refraction for orthogonal linear polarizations of the light. The phase velocity of a wave in a birefringent medium thus depends on the polarization of the wave. Fibers may exhibit birefringence. *See also: REFRACTIVE INDEX (of a medium).*

bit *Acronym for BINARY DIGIT. See: ABORT; ADDED BIT; AVERAGE BLOCK LENGTH; BLOCK; BYTE; CHARACTER-COUNT AND BIT-COUNT INTEGRITY; DATA SIGNALING RATE; BIT; DE-STUFFING; DIBIT; ERROR BURST; FRAME; INFORMATION BIT; MULTIPLEX AGGREGATE BIT RATE; PARITY BIT.*

bit-by-bit asynchronous operation A mode of operation in which rapid manual, semiautomatic, or automatic shifts in the data modulation rate are accomplished by gating or slewing the clock modulation rate. The equipment may be operated at 50 b/s one moment and at 1200 b/s the next moment, etc.

bit count integrity (BCI) *See: CHARACTERCOUNT AND BIT-COUNT INTEGRITY.*

biternary transmission A method of digital transmission in which two binary pulse trains are combined for transmission over a system in which the available bandwidth is only sufficient for transmission of one of the two pulse trains when in binary form. The biternary signal is generated from two synchronous binary signals operating at the same bit rate. The two binary signals are adjusted in time to have a relative time difference of one-half the binary interval and are combined by linear addition to form the biternary signal. Each biternary signal element can assume any one of three possible states, i.e., $+1$, 0, or -1. Each biternary signaling element contains information on the state of the two binary signaling elements as defined in the following truth table:

B1	B2	BITERNARY
0	0	-1
0	1	0
1	0	0
1	1	$+1$

The method of addition of B1 and B2 as described above does not permit the biternary signal to change from -1 to $+1$ or $+1$ to -1 without an intermediate biternary signal of 0. Since there is half a unit interval time difference between the binary signals B1 and B2, only one of them can change its state during the biternary unit interval. This makes it possible in the decoding process to ascertain the state of the binary signal that has not changed its state and thus avoid ambiguity in decoding a biternary signal of 0.

bit error rate (BER) The number of erroneous bits divided by the total number of bits over some stipulated period of time.

Note: Two examples of bit error rate are: a) transmission BER-number of erroneous bits received divided by the total number of bits transmitted; and b) information BER-number of erroneous decoded (corrected) bits divided by the total number of decoded (corrected) bits. The BER is usually expressed as a number and a power of 10, e.g., 2.5 erroneous bits out of 100,000 bits transmitted would be 2.5 in 10^{-5} or 2.5 x 10^{-5}. *See also: ERROR.*

bit inversion The deliberate or fortuitous changing of the state of a bit to the opposite state. *See also: CHARACTER-COUNT AND BIT-COUNT INTEGRITY.*

bitoric lens A lens both surfaces of which are ground and polished in a cylindrical form or in a toroidal shape.

bit pairing The practice of establishing, within a code set, a number of subsets (of two characters each) that have an identical bit representation except for the state of a specified bit. For example, in the International Telegraph Alphabet Number 5 and the American National Standard Code for Information Interchange (ASCII), the upper case letters are related to their respective lower case letters by the state of bit six.

bit-rate-length product For an optical fiber or cable, the produce of bit rate the fiber or cable is able to handle and the length for tolerable dispersion at the bit rate. The product usually stated in units of Mb–km/s.

Note: Typical bit-rate-length products for graded index fibers with a numerical aperture (NA) of 0.2 is 100 Mb–km/s for research fibers and 200 Mb–km/s for production fibers; plastic clad fibers with an (NA) of 0.25 are 30 Mb–km/s for both research and production fibers. The product is a good measure of fiber performance in terms of transmission capability.

bit-sequence independence In a digital path or digital section of a PCM system operating at a specified bit rate, a measure of the capability to permit any sequence of bits at that rate, or the equivalent, to be transmitted.

Note: Practical transmission systems which are not completely bit-sequence independent may be described as quasi bit-sequence independent. In such cases the limitations should be clearly stated.

bit sequential A type of transmission in which the elements of a signal are successive in time.

bits per second (b/s) The number of bits passing a point per second.

Note: Values of modulation rate in baud and in bits per second are numerically the same if, and only if, all of the three following conditions are met: 1. All pulses (bits) are the same length. 2. All pulses (bits) are equal to the unit interval, the time element between the same two significant instants of adjacent pulses. 3. Binary operation is used. In N-ary operation, b/s equals modulation rate in baud multiplied by the logarithm to the base 2 of N, where N is the number of distinct states of amplitude, phase, etc., used in the digital modulation process. *See also: BAUD; DATA SIGNALING RATE.*

bit stepped Operational control of digital equipment in which a device is stepped one bit at a time at the applicable modulation rate.

bit stream transmission The transmission of characters at fixed time intervals without stop and start elements, the bits that make up the characters following each other in sequence without interruption.

bit string A linear sequence of bits.

bit stuffing A synchronization method used in time division multiplexing to handle received bit streams over which the multiplexer clock has no control. The modulation rate of the multiplexer is made somewhat higher than the sum of the incoming rates, by "stuffing in" additional pulses. Each incoming signal has enough pulses added to it to give it a rate compatible with the multiplexer clock. The number of pulses stuffed on the individual incoming signals is varied to take care of variations in their rates. The multiplexer inserts a code as part of the stuffing to tell the demultiplexer what pulses to "de-stuff" so that the signal delivered to the outgoing circuit has exactly the same number of pulses as that received. *Synonyms: POSITIVE JUSTIFICATION; PULSE STUFFING; STUFFING. See also: DESTUFFING; MULTIPLEX; NEGATIVE ACKNOWLEDGE CHARACTER; NOMINAL BIT STUFFING RATE.*

bit stuffing rate *See: NOMINAL BIT STUFFING RATE.*

bit synchronous operation A mode of

15

operation in which data circuit-terminating equipment (DCE), data terminal equipment (DTE), and transmitting circuits are all operated in bit synchronism with an accurate clocking system, i.e., they are all synchronous with the clock. Clock timing is delivered at twice the modulation rate, and one bit is released during one clock cycle.

Note: Bit synchronous operation is sometimes erroneously referred to as Digital synchronization. *See also: CLOCK; DATA TERMINAL EQUIPMENT.*

blackbody   A totally absorbing body (which reflects no radiation).

Note: In thermal equilibrium, a blackbody absorbs and radiates at the same rate; the radiation will just equal absorption when thermal equilibrium is maintained. *See also: EMISSIVITY.*

blacknoise   In a spectrum of electromagnetic wave frequencies, a frequency spectrum of predominantly zero power level at all frequencies except for a few narrow bands or spikes, such as might be obtained when scanning a black area in facsimile transmission systems and there are a few white spots or speckles on the surface.

blemish   An area, in a fiber or fiber bundle, that has a reduced light transmission capability, i.e., increased attenuation, due to defective or broken fibers, foreign substances, or other spoilage.

block   1. A group of bits, or N-ary digits, transmitted as a unit. An encoding procedure is generally applied to the group of bits or N-ary digits for error-control purposes. 2. A string of records, a string of words, or a character string, formed for technical or logic reasons to be treated as an entity. 3. A set of things, such as words, characters, or digits, handled as a unit. 4. A collection of contiguous records recorded as a unit. Blocks are separated by interblock gaps and each block may contain one or more records. *See also: ADDED BLOCK; ADDED-BLOCK PROBABILITY; AVERAGE BLOCK LENGTH; BLOCK TRANSFER TIME; DELETED BLOCK; DELIVERED OVERHEAD BLOCK; DIGITAL BLOCK; END-OF-TRANSMISSION-BLOCK CHARACTER; ERRONEOUS BLOCK; INCORRECT BLOCK; LOST BLOCK; MAXIMUM BLOCK TRANSFER TIME; MISDELIVERED BLOCK; SUCCESSFUL BLOCK DELIVERY; SUCCESSFUL BLOCK*

TRANSFER; TIME BLOCK; TIME CODE; USER INFORMATION BLOCK.

blocking cement   An adhesive used to bond optical elements to blocking tools. *Note:* It is usually a thermoplastic material such as resin, beeswax, pitch, or shellac.

blue noise   In a spectrum of electromagnetic wave frequencies, a region in which the spectral density is proportional to the frequency (sloped), rather than independent of frequency (flat), as in white noise that is more of a uniformly distributed constant amplitude frequency spectrum.

body   *See: BLACKBODY.*

BOG laser   *See: BURIED-OPTICAL-GUIDE DIODE LASER.*

bolometer   A device for measuring radiant energy by measuring the changes in resistance of a temperature sensitive device exposed to radiation. *See also: RADIANT ENERGY; RADIOMETRY.*

Boltzmann's constant (k)   The number k that relates the average energy of a molecule to the absolute temperature of the environment. k is approximately 1.38 x 10^{-23} joules/kelvin.

Bouger's law   In the transmission of electromagnetic radiation through a material medium the attenuation, reduction, decay, or diminution of electromagnetic field intensity or power density is an exponential decay function of the product of a constant coefficient dependent upon the material and the thickness given by the relationship

$$I = I_0 \, e^{-AX}$$

where I is the intensity at distance X, I_0 is the intensity at $X = 0$, and A is a material constant coefficient that depends upon the scattering and absorptive properties of the medium. If only absorption takes place, A is the spectral absorption coefficient and is a function of wavelength; if only scattering takes place, it is the scattering coefficient; if both absorption and scattering occur, it is the extinction coefficient being then the sum of the absorption and scattering coefficients. *See also: BEER'S LAW; LAMBERT'S LAW.*

boundary-layer photocell   *See: PHOTOVOLTAIC CELL.*

bound mode   In an optical waveguide, a mode whose field decays monotonically in the transverse direction everywhere external to the core and which does not lose power to radiation. Specifically, a mode for which

$$n(a)k \leq \beta \leq n(0)k$$

where β is the imaginary part (phase constant) of the axial propagation constant, n(a) is the refractive index at r=a, the core radius, n(0) is the refractive index at r=0, k is the free-space wavenumber, $2\pi/\lambda$, and λ is the wavelength. Bound modes correspond to guided rays in the terminology of geometric optics. *Note:* Except in a monomode fiber, the power in bound modes is predominantly contained in the core of the fiber. *Synonyms: GUIDED MODE; TRAPPED MODE. See also: CLADDING MODE; GUIDED RAY; LEAKY MODE; MODE; NORMALIZED FREQUENCY; UNBOUND MODE.*

bound ray *Synonym: GUIDED RAY.*

branched cable A multiple-wire, multiple-fiber, or multiple-bundle cable that contains one or more breakouts, divergencies, i.e., branches.

branching repeater A repeater with two or more outputs for each input.

breakdown voltage Voltage value, while a diode is reverse biased, at which avalanche breakdown occurs (i.e., for a small voltage increase the drive current tends to infinity). Symbol: V_B. The breakdown voltage is an important characteristic of APDs. V_B is roughly inversely proportional to the carrier concentration on the lightly doped side of the p-n junction. In diodes containing defects, local carrier multiplication occurs at microplasma sites which decreases V_B below the theoretical value. Junction breakdown can also occur due to tunnelling. *Note:* For example see S. M. Sze, *Physics of Semiconductor Devices,* 2nd edition, John Wiley & Sons, 1981.

breakout point The point where a branch meets, merges, or joins with or diverges from the main cable or harness run. *Note:* The convergence or divergence is the breakout point of the fiber.

Brewster's angle For light incident on a plane boundary between two regions having different refractive indices, that angle of incidence at which the reflectance is zero for light that has its electric field vector in the plane defined by the direction of propagation and the normal to the surface. For propagation from medium 1 to medium 2, Brewster's angle is

$$\arctan (n_2/n_1).$$

See also: ANGLE OF INCIDENCE; REFLECTANCE; REFRACTIVE INDEX (of a medium).

Brewster's law When an electromagnetic wave is incident upon a surface, and the angle between the refracted and reflected ray is 90°, maximum polarization occurs in both rays, the reflected ray having its maximum polarization in a direction normal to the plane of incidence, and the refracted ray having its maximum polarization in the plane of incidence.

brightness An attribute of visual perception, in accordance with which a source appears to emit more or less light; obsolete. *Note:* 1. Usage should be restricted to nonquantitative reference to physiological sensations and perceptions of light. 2. "Brightness" was formerly used as a synonym for the photometric term "luminance" and (incorrectly) for the radiometric term "radiance." *See also: RADIANCE; RADIOMETRY.*

brightness conservation *See: RADIANCE CONSERVATION.*

Brillouin scattering Inelastic scattering of photons by phonons. The part of the scattered radiation in a medium (liquid or solid) whose frequency is shifted from the incident radiation by amounts corresponding to the energy of phonons (lattice vibrations) created or annihilated via scattering (i.e., during the process the incident photon causes the excitation or de-excitation of a lattice vibration). Of importance in optical fibers at high optical power densities, and only above a threshold (i.e., onset of stimulated Brillouin scattering).

broadband system *Synonym: WIDEBAND SYSTEM.*

broadcast operation The transmission of information so that it may be received by stations that usually make no acknowledgment.

broadcast repeater A repeater connecting several channels, one incoming and the others outgoing.

buffer *See: FIBER BUFFER.*

bulk degradation Diode laser or LED degradation due to stress or poor material quality (rapid bulk degradation) or to point-defect generation and propagation (slow bulk degradation). Slow bulk degradation ultimately determines the devices' lifetime, if mirror-facet degradation and metallization failure are eliminated.

bulk material absorption In fiber optics, the lightwave power absorption that occurs per unit volume of the basic material used to form an optical fiber either core, cladding or jacket.
Note: Measurement is made of bulk material absorption prior to use in forming optical waveguides. Absorption is usually expressed in dB/km, i.e., as an attenuation.

bulk material scattering In fiber optics, the lightwave per unit power that is scattered per unit volume of the basic material used to form an optical fiber, either core, cladding, or jacket.
Note: Measurement is made of bulk material scattering prior to use in forming optical waveguides. Scattering follows a Rayleigh distribution, characteristic of a medium whose refractive index fluctuates over small distances compared to the wavelength of the incident light. Scattering losses are usually expressed in dB/km.

bunched frame-alignment signal A frame-alignment signal in which the signal elements occupy consecutive digit positions. *See also: DISTRIBUTED FRAME-ALIGNMENT SIGNAL.*

bundle *See: FIBER BUNDLE.*

bundle cable *See: MULTIPLE-BUNDLE CABLE; MULTI-CHANNEL BUNDLE CABLE; SINGLE-CHANNEL SINGLE-BUNDLE CABLE.*

bundle cable assembly *See: MULTIPLE-BUNDLE CABLE ASSEMBLY.*

bundle jacket The outer protective covering applied over a bundle of optical fibers. *See also: CLADDING.*

bundle resolving power The ability of a coherent optical fiber bundle to transmit the details of an image. Usually stated in lines per millimeter.

buried crescent (BC) diode laser Mode-stabilized laser grown by one-step liquid-phase epitaxy above a channel etched into the substrate. The active layer is of convex-lens shape above the channel and tapers down to zero thickness on either side of the channel. Typi-cal reliable power capability: 3-5 mW/facet cw.
Note: See for example Oomura, *et al., IEEE J. Quantum Electron.,* vol. 17, pp. 646–650, May 1981.

buried-facet diode laser A diode laser for which the mirror facets are partially or totally composed of a high-resistivity material transparent to the lasing emission and deposited after growth and preferential etching of a light-emitting structure. *See also: NONABSORBING-MIRROR STRUCTURES.*

buried heterostructure diode laser A mode-stabilized diode laser structure realized by embedding a mesa of planar DH material in high-resistivity material. The embedding material provides lateral carrier and optical confinement. The most popular mode-stabilized device for long-wavelength operation.

buried heterostructure large-optical-cavity laser A mode-stabilized diode laser structure realized by embedding a mesa of planar LOC material in high-resistivity material.
Note: See for example D. Botez, *J. Opt. Commun.,* vol. 1, pp. 42–50, Nov. 1980.

buried-optical-guide laser *See: BURIED HETEROSTRUCTURE LARGE-OPTICAL-CAVITY LASER.*
Note: For example see N. Chinone, *et al., Appl. Phys. Lett.,* vol. 35, pp. 513–516, Oct. 1979.

Burrus LED *See: SURFACE-EMITTING LED.*

burst 1. In data communications, a sequence of signals counted as a unit in accordance with some specific criterion or measure. *See also: ERROR BURST.* 2. To separate continuous-form paper into discrete sheets. 3. To separate multipart paper.

burst isochronous Deprecated. *See: ISOCHRONOUS BURST TRANSMISSION.*

burst transmission A method of operating a data network by interrupting, at intervals, the data being transmitted.
Note: The method enables communication between data terminal equipment and a data network operating at dissimilar data signaling rates.

bus One or more conductors that serve as a common connection for a related group of devices. *See also: DATA BUS.*

bus coupler *See: DATA-BUS COUPLER.*

byte A sequence of consecutive bits, usually shorter than a word, operated on as a unit.

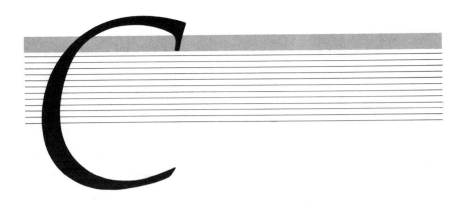

cable assembly *See: MULTIFIBER CABLE; OPTICAL CABLE ASSEMBLY.*

cable core The portion of a cable inside of a common covering.

cable jacket The outer protective covering applied over the internal cable elements.

cable run The portion of a branched cable or harness where the cross-sectional area of the cable or harness is the largest. *Synonym: HARNESS RUN.*

call-second A unit of communication traffic, such that one call-second may be defined as one user making one call of one second duration.

Note: One user making two 75-second calls or two users, each making a 75-second call apiece, produce the same 150 call-seconds of traffic. Since a larger unit than the call-second is generally needed, the CCS (hundred-call-second) was introduced.

RELATIONSHIPS

3600 call-seconds = 36 CCS = 1 call hour. 3600 call-seconds per hour = 36 CCS per hour = 1 call hour per hour = 1 Erlang = 1 traffic unit. *See also: ERLANG; TRAFFIC INTENSITY.*

calspar *See: ICELAND SPAR.*

canada balsam An adhesive used to cement optical elements.

candela The luminous intensity of 1/600,000 of a square meter of a blackbody radiator at the temperature of solidification of platinum, 2045 kelvins. One candela emits 4π lumens of light flux.

candlepower A unit of measure of the illuminating power of any light source, equal to the number of candles of the source of light. A flux density of one lumen of luminous flux per steradian of solid angle measured from the source is produced by a point source of one candela emitting equally in all directions.

capacity *See CHANNEL CAPACITY; TRAFFIC CAPACITY.*

capping layer The last layer grown in a heterostructure to provide good electrical contact between the semiconductor body and the contact metallization.

Note: For AlGaAs lasers the capping layer is GaAs, while for InGaAsP lasers the capping layer is InGaAsP. Capping layers are usually heavily doped for good electrical contact.

capture effect An effect associated with the reception of frequency modulated signals in which, if two signals are received on the same frequency, only the stronger of the two will appear in the output. The complete suppression of the weaker carrier occurs at the receiver limiter, where it is treated as noise and rejected.

carrier 1. A wave suitable for modulation by an information-bearing signal to be transmitted over a communication system. 2. An unmodulated emission.

Note: The carrier is usually a sinusoidal wave or a recurring series of pulses. *Synonym: CARRIER WAVE. See also: CENTER FREQUENCY; DOUBLE-SIDEBAND SUPPRESSED CARRIER TRANSMISSION; SINGLE-SIDEBAND SUPPRESSED CARRIER; SUBCARRIER.*

carrier concentration The volume concentration of electrons or holes injected into the active region of a laser or LED at a given current density.

carrier dropout A short-duration carrier signal loss.

carrier frequency 1. The frequency of a carrier wave. 2. A frequency capable of being modulated or impressed with a second (information carrying) signal.

Note: In frequency modulation, the carrier frequency is also referred to as the center frequency.

carrier leak The carrier remaining after carrier suppression in a suppressed carrier transmission system.

carrier leakage Loss of carriers out of the active region of a laser or LED by drift or diffusion. Carrier leakage by diffusion is a strong function of temperature and can severely impair diode performance at high temperatures.

carrier level The power of a carrier signal at a particular point in a system, expressed in decibels in relation to some reference level.

carrier lifetime The average decay time for a given concentration of excess carriers that recombine both radiatively and nonradiatively. *See also: RADIATIVE CARRIER LIFETIME.*

carrier multiplex *Synonym: FREQUENCY DIVISION MULTIPLEX.*

carrier noise level The noise level resulting from undesired variations of a carrier in the absence of any intended modulation. *Synonym: RESIDUAL MODULATION. See also: NOISE.*

carrier power The average power supplied to the antenna transmission line by a radio transmitter during one radio frequency cycle under conditions of no modulation.
Note: This definition does not apply to pulse modulated emissions or FSK.

carrier shift 1. A method of keying a radio carrier for transmitting binary data or teletypewriter signals which consists in shifting the carrier frequency in one direction for a marking signal and in the opposite direction for a spacing signal. 2. A condition resulting from imperfect modulation whereby the positive and negative excursions of the envelope pattern are unequal, thus effecting a change in the power associated with the carrier. There can be positive or negative carrier shift.

carrier-to-noise ratio (CNR) The ratio, in decibels, of the value of the carrier to that of the noise in the receiver IF bandwidth before any nonlinear process such as amplitude limiting and detection.

carrier-to-receiver noise density In satellite communications, the ratio, expressed in dB, of the received carrier power (C) to the received noise power density (kT), where k is Boltzmann's constant and T is the receiver system noise temperature in kelvin. The ratio is also referred to as C/kT.

carrier wave *Synonym: CARRIER.*

cartesian lens A lens, one surface of which is a cartesian oval, thus producing an aplanatic condition.

case shift 1. In data equipment, the change from letters to numerals, or vice versa. 2. In typewriting or type setting, the change from lower case letters to upper case letters, or vice versa.

catastrophic degradation In laser diodes, the sudden reduction in optical power output, excessive stress, current transients, or thermal runaway.
Note: The critical facet; damage is measured in watts per centimeter square of emitting area. Catastrophic degradation has occurred in solid-state (semiconductor) lasers of all types. *See also: GRADUAL DEGRADATION; CATASTROPHIC OPTICAL DAMAGE.*

catastrophic optical damage (COD) Laser diode irreversible damage caused at very high optical flux densities at the mirror facet (i.e., catastrophic facet damage) or, in nonabsorbing-mirror devices, in the bulk (i.e., catastrophic bulk damage). The COD power density level is a function of pulse width, type of facet-passivation coating used and material used. COD levels in AlGaAs lasers are 4–5 times higher than the power levels for reliable operation.

cavity *See: OPTICAL CAVITY.*

cavity diode *See: LARGE OPTICAL-CAVITY DIODE.*

CCH *Acronym for CONNECTIONS PER CIRCUIT HOUR.*

CCM laser *See: CLOSE-CONFINEMENT MESA LASER.*

CCS *Acronym for HUNDRED-CALL-SECOND. See CALL-SECOND.*

CDH laser *See: CONSTRICTED DOUBLE-HETEROJUNCTION DIODE LASER.*

CDH-LOC laser *See: CONSTRICTED DOUBLE-HETEROJUNCTION LARGE-OPTICAL-CAVITY DIODE LASER.*

CDM *Acronym for Color-division multiplexing.*

CDMA *Acronym for CODE-DIVISION MULTIPLE ACCESS.*

cell *See: KERR CELL; PHOTOEMISSIVE CELL; PHOTOVOLTAIC CELL; POCKEL CELL.*

cement In optics, an adhesive used to bond optical elements together, or to bond optical elements to holding devices.
Note: Three general types of cement used in the optical industry are blocking ce-

ments; mounting cements; and optical cements. *See also: BLOCKING CEMENT; MOUNTING CEMENT; OPTICAL CEMENT; THERMOPLASTIC CEMENT; THERMOSETTING CEMENT.*

cemented doublet In optics, a compound lens of two elements cemented together over their contiguous surfaces.

center frequency 1. In frequency modulation, the resting frequency or initial frequency of the carrier before modulation. 2. In facsimile, the frequency midway between picture black and picture white frequencies. *See also: CARRIER.*

center sampling A method of sampling a digital data stream at the center of each signal element.

central strength-member optical cable A cable containing optical fibers that are on the outside of, or wrapped around, a high tensile-strength material, such as stranded steel, nylon, or other material, with crush-resistant jacketing (sheathing) on the outside of the cable. *See also: PERIPHERAL STRENGTH-MEMBER OPTICAL CABLE.*

channel 1. A single unidirectional or bidirectional path for transmitting or receiving, or both, of electrical signals, usually in distinction from other parallel paths. 2. A single path provided from a transmission medium either by physical separation (e.g., multi-pair cable) or by electrical separation (e.g., frequency or time division multiplexing). 3. A connection between initiating and terminating nodes of a circuit. 4. A path along which signals can be sent, e.g., data channel, output channel. 5. The portion of a storage medium that is accessible to a given reading or writing station, e.g., track, band. 6. In information theory, that part of a communications system that connects the message source with the message sink. *See also: ADAPTIVE CHANNEL ALLOCATION; ADJACENT-CHANNEL INTERFERENCE; AUXILIARY CHANNEL; BACKWARD CHANNEL; CHANNELIZATION; DATA TRANSMISSION CIRCUIT; DROP CHANNEL; FORWARD CHANNEL; INFORMATION BEARER CHANNEL; INFORMATION CHANNEL; NONSYNCHRONOUS DATA TRANSMISSION CHANNEL; ONE-WAY-ONLY CHANNEL; PRIMARY CHANNEL; RADIO CHANNEL; SECONDARY CHANNEL; STATISTICAL MULTIPLEXING; SYMMETRICAL CHANNEL; TRANSMIS-*

SION CHANNEL; VOICE FREQUENCY CHANNEL.

channel-associated signaling Signaling in which the signals necessary for the traffic carried by a single channel are transmitted in the channel itself or in a signaling channel permanently associated with it. *See also: INBAND SIGNALING.*

channel bank A part of a carrier-multiplex terminal that performs the first step of modulation. It multiplexes a group of channels into a higher frequency band and, conversely, demultiplexes the higher frequency band into individual channels. *See also: WIDEBAND SYSTEM.*

channel bundle cable *See: MULTI-CHANNEL BUNDLE CABLE.*

channel cable *See: MULTI-CHANNEL CABLE.*

channel capacity A measure of the maximum possible information rate through a channel, subject to specified constraints.

channel gate A device for connecting a channel to a highway, or a highway to a channel, at specified times.

channelization The method of using a single wideband facility to transmit many relatively narrow bandwidth channels by subdividing the wideband channel. *See also: CHANNEL; WIDEBAND SYSTEM.*

channel-narrow-stripe (CNS) diode laser Mode-stabilized diode laser structure grown by one-step liquid phase epitaxy over a channel in the substrate such that the active layer assumes a convex-lens-like shape above the channel. Current confinement is provided by narrow (2–4 μm) contact stripes. Reliable power capability is $\cong$5 mW/facet.

Note: For example see P. A. Kirkby, *Electron. Lett.,* vol. 15, pp. 824–825, 1979.

channel noise level 1. The ratio of the channel noise at any point in a transmission system to some arbitrary amount of circuit noise chosen as a reference. This ratio is usually expressed in decibels above reference noise, abbreviated dBrn, signifying the reading of a circuit noise meter, or in adjusted decibels, abbreviated dBa, signifying circuit noise meter reading adjusted to represent an interfering effect under specified conditions. 2. The noise power density spectrum in the frequency range of interest. 3. The average noise power in the frequency range of interest. 4. The indication on a specified instrument. The characteristics of the instrument are determined by the type of noise to be measured and the application of the results thereof. *See also: NOISE; SIGNAL-TO-NOISE RATIO.*

channel packing A technique for maximizing the utilization of voice frequency channels used for data transmission by multiplexing a number of lower speed data signals into a single higher speed data stream for transmission on a single voice frequency channel.

channel reliability The percent of time a channel was available for use in a specific direction during a specified period of scheduled availability. Channel reliability, CR, is given in percent by:

$$CR = 100\ (1 - TO/TS) = 100\ TA/TS$$

where TO is the channel total outage time, TS is the channel total scheduled time, and TA is the channel total available time. *See also: CIRCUIT RELIABILITY.*

channel single-bundle cable *See: SINGLE-CHANNEL SINGLE-BUNDLE CABLE.*

channel single-fiber cable *See: MULTI-CHANNEL SINGLE-FIBER CABLE; SINGLE-CHANNEL SINGLE-FIBER CABLE.*

channel-substrate-planar (CSP) diode laser Mode-stabilized diode laser structure grown by one-step liquid-phase epitaxy over a channel in the substrate such that the active layer is flat and of constant thickness and the first confinement layer fully fills the channel. Typical reliable power capability: 10–15 mW/facet cw.
Note: See for example Aiki, *et al., IEEE J. Quantum Electron.,* vol. 14, pp. 89–97, Feb. 1978.

channel supergroup *See GROUP.*

channel time slot A time slot starting at a particular instant in a frame and allocated to a channel for transmitting a character, in-slot signal, or other data.
Note: Where appropriate, a modifier may be added, for example "telephone channel time slot."

character A letter, digit, or other symbol that is used as part of the organization, control, or representation of data. A character is often in the form of a spatial arrangement of adjacent or connected strokes. *See also: ACKNOWLEDGE CHARACTER; BLOCK CHECK CHAR-*

ACTER; CALL CONTROL CHARACTER; CODE CHARACTER; CODED CHARACTER SET; COMPUTER WORD; CONTROL CHARACTER; DATA COMMUNICATION CONTROL CHARACTER; DATA LINK ESCAPE CHARACTER; DIGIT; END-OF-SELECTION CHARACTER; END-OF-TEXT CHARACTER; END-OF-TRANSMISSION-BLOCK CHARACTER; END-OF-TRANSMISSION CHARACTER; ENQUIRY CHARACTER; FACILITY REQUEST SEPARATOR CHARACTER; HANDSHAKING; IDLE CHARACTER; NEGATIVE ACKNOWLEDGE CHARACTER; START-OF-HEADING CHARACTER; START-OF-TEXT CHARACTER; SYNCHRONIZATION BIT.

character-count and bit-count integrity The preservation of the precise number of characters, or bits, that are originated in a message (in the case of message communication) or per unit time (in the case of a user-to-user connection).
Note: Not to be confused with bit integrity or character integrity which require that the characters or bits delivered are, in fact, as they were originated. *See also: ADDED BIT; BIT; BIT INVERSION; DELETED BIT; DIGITAL ERROR.*

character interval The total number of unit intervals (including synchronizing, information, error checking, or control bits) required to transmit any given character in any given communications system. Extra signals which are not associated with individual characters are not included.
Note: An example of an extra signal that is excluded in the above definition is any additional time added between the end of the stop element and the beginning of the next start element as a result of a speed change, buffering, etc. This additional time is defined as a part of the intercharacter interval. *See also: INTERCHARACTER INTERVAL.*

characteristic distortion Distortion caused by transients which, as a result of modulation, are present in the transmission channel. Its effects are not consistent; its influence upon a given transition is to some degree dependent upon the remnants of transients affecting previous signal elements.

characteristic frequency A frequency which can be easily identified and measured in a given emission.

characteristic impedance The impedance that an infinitely long transmission line would present at its input terminals. A line will appear to be infinitely long if terminated in its characteristic impedance.

Note: It is recommended that this term be applied only to lines having approximate electrical uniformity. For other lines or structures the corresponding term is *ITERATIVE IMPEDANCE.*

character set 1. A finite set of different characters upon which agreement has been reached and that is considered complete for some purpose, e.g., each of the character sets in ISO Recommendation R646 "6- and 7-bit Coded Character Sets for Information Processing Interchange." 2. An ordered set of unique representations called characters, e.g., the 26 letters of the English alphabet, Boolean 0 and 1, and the 128 ASCII characters.

charge *See: ELECTRONIC CHARGE.*

chemical vapor deposition (CVD) technique A process in which deposits are produced by heterogeneous gassolid and gas-liquid chemical reactions at the surface of a substrate.

Note. The CVD method is often used in fabricating óptical waveguide preforms by causing gaseous materials to react and deposit glass oxides. Typical starting chemicals include volatile compounds of silicon, germanium, phosphorus, and boron, which form corresponding oxides after heating with oxygen or other gases. Depending upon its type, the preform may be processed further in preparation for pulling into an optical fiber. *See also: PREFORM.*

chemical vapor phase oxidation process (CVPO) A process for the production of low-loss (less than 10 dB/km), high bandwidth (greater than 300 MHz-km), multimode, graded index (GI) optical fiber, involving either the inside vapor phase oxidation (IVPO) process, the outside vapor phase oxidation (OVPO) process, the modified chemical vapor deposition (MCVD) process, the plasma-activated chemical vapor deposition (PCVD) process, or the axial vapor phase oxidation (AVPO) process, or a combination or variation of these, by soot deposition on a glass substrate followed by oxidation and drawing of the fiber. *Synonym: SOOT PROCESS.*

chicken wire An optical fiber blemish that appears as a grid of lines along fiber boundaries in a multifiber bundle.

chief ray The central ray of a bundle of rays of light.

chip *See: FLIP CHIP.*

chirping A rapid change (as opposed to long-term drift) of the emission wavelength of an optical source. Chirping is most often observed in pulsed operation of a source.

chromatic aberration Image imperfection caused by light of different wavelengths following different paths through an optical system due to dispersion caused by the optical elements of the system.

chromatic dispersion Redundant synonym for dispersion.

chromatic resolving power The ability of an instrument to separate two electromagnetic-wave wavelengths, equal to the ratio of the shorter wavelength divided by the difference between the wavelengths.

Note: Resolving power normally refers to the ability of optical components to separate two or more object points close together. *See also: GRATING CHROMATIC RESOLVING POWER; PRISM CHROMATIC RESOLVING POWER.*

circuit 1. The complete path between two end-terminals over which one-way or two-way communications may be provided. 2. An electronic path between two or more points capable of providing a number of channels. 3. A number of conductors connected together for the purpose of carrying an electrical current. *See also: CIRCUITRY; CLOSED CIRCUIT; COMMON USER CIRCUIT; CONDITIONED CIRCUIT; CONDITIONED VOICE GRADE CIRCUIT; CONTINUITY CHECK; CORD CIRCUIT; DATA CIRCUIT CONNECTION; DECISION CIRCUIT; DEDICATED CIRCUIT; DRIVING CIRCUIT; DUPLEX CIRCUIT; FOURWIRE CIRCUIT; GROUND-RETURN CIRCUIT; HALF-DUPLEX CIRCUIT; INTEGRATED CIRCUIT; INTERCHANGE CIRCUIT; INVERTER; LABEL; LIMITER CIRCUIT; METALLIC CIRCUIT; MULTIPOINT CIRCUIT; NODE; OPEN CIRCUIT; INTEGRATED OPTICAL CIRCUIT: ORDERWIRE CIRCUITS; PERMANENT VIRTUAL CIRCUIT; PHANTOM CIRCUIT; PILOT MAKE BUSY CIRCUIT; REFERENCE CIRCUIT; RINGDOWN CIRCUIT; SIDE CIRCUIT; SIM*

PLEX CIRCUIT; STEADY STATE CONDITION; TRANSMISSION CHANNEL; TWO-WIRE CIRCUIT; UNBALANCED WIRE CIRCUIT; VIRTUAL CIRCUIT.

circuit filter-coupler-switch-modulator See: INTEGRATED-OPTICAL CIRCUIT FILTER-COUPLER-SWITCH-MODULATOR.

circuit noise level At any point in a transmission system, the ratio of the circuit noise at that point to some arbitrary amount of circuit noise (usually expressed in dBm) chosen as a reference. The ratio is usually expressed in decibels above reference noise, abbreviated dBrn, signifying the reading of a circuit noise meter, or in dBrn adjusted, abbreviated dBa, signifying circuit noise meter reading adjusted to represent an interfering effect under specified conditions. See also: dBrnC; NOISE.

circuit reliability The percentage of time a circuit was available to the user during a specified period of scheduled availability. Circuit reliability, CR, is given in percent by:

$$CR = 100 \, (1 - TO/TS) = 100 \, TA/TS$$

where TO is the circuit total outage time, TS is the circuit total scheduled time, and TA is the circuit total available time. Synonym: TIME AVAILABILITY. See also: CHANNEL RELIABILITY.

circuitry A complex of circuits describing interconnection within or between systems. See also: CIRCUIT.

circuit-switched data transmission service A service requiring the establishment of a circuit-switched data connection before data can be transferred between data terminal equipment.

circuit switching A method of handling traffic through a switching center, either from local users or from other switching centers, whereby a connection is established between the calling and called stations. See also: MESSAGE SWITCHING; SWITCHING SYSTEM.

circuit switching center (CSC) A communications-electronics complex of circuits, equipment, and supporting facilities used for establishing a connection between two compatible subscribers.

circuit switching unit (CSU) That equipment used for directly connecting two compatible data terminals for end-to-end data exchanges. Also used to connect a data terminal to a store-and-forward switch.

circular birefringence See: OPTICALLY ACTIVE MATERIAL.

circulator A passive junction of three or more ports in which the ports can be listed in such an order that when power is fed into any port it is transferred to the next port on the list, the first port being counted as following the last in order.

cladded fiber See: DOPED-SILICA CLADDED FIBER.

cladding The dielectric material surrounding the core of an optical waveguide. See also: CORE; NORMALIZED FREQUENCY; OPTICAL WAVEGUIDE; TOLERANCE FIELD.

cladding center The center of the circle that circumscribes the outer surface of the homogeneous cladding, as defined under Tolerance field. See also: CLADDING; TOLERANCE FIELD.

cladding diameter The length of the longest chord that passes through the fiber axis and connects two points on the periphery of the homogeneous cladding. See also: CLADDING; CORE DIAMETER; TOLERANCE FIELD.

cladding-guided mode In an optical waveguide, a transmission mode supported by the cladding, i.e. a mode in addition to the modes supported by the core material.
Note: Cladding-guided modes are usually attenuated by absorption by using lossy cladding media to prevent reconversion of energy to core-guided modes and thus reduce dispersion.

cladding layer See: CONFINEMENT LAYER.

cladding mode A mode that is confined by virtue of a lower index medium surrounding the cladding. Cladding modes correspond to cladding rays in the terminology of geometric optics. See also: BOUND MODE; CLADDING RAY; LEAKY MODE; MODE; UNBOUND MODE.

cladding mode stripper A device that encourages the conversion of cladding modes to radiation modes; as a result, the cladding modes are stripped from the fiber. Often a material having a refractive index equal to or greater than that of the waveguide cladding. See also: CLADDING; CLADDING MODE.

cladding ray In an optical waveguide, a ray that is confined to the core and cladding by virtue of reflection from the outer surface of the cladding. Cladding rays correspond to cladding modes in the terminology of mode descriptors. *See also: CLADDING MODE; GUIDED RAY; LEAKY RAY.*

clad silica fiber *See: LOW-LOSS FEP-CLAD SILICA FIBER; PLASTIC-CLAD SILICA FIBER.*

clipper A circuit or device that limits the instantaneous output signal amplitude to a predetermined maximum value, regardless of the amplitude of the input signal. *See also: COMPANDOR; COMPRESSOR; EXPANDOR; LIMITER; PEAK LIMITING.*

clock 1. A reference source of timing information for equipment, machines, or systems. 2. Equipment providing a time base used in a transmission system to control the timing of certain functions such as the control of the duration of signal elements or the sampling rate. *See also: BIT SYNCHRONOUS OPERATION; DoD MASTER CLOCK; EQUIPMENT CLOCK; MASTER CLOCK; NONSYNCHRONOUS NETWORK; REFERENCE CLOCK; SINGLE-ENDED SYNCHRONIZATION; STATION CLOCK; SYNCHRONOUS NETWORK; TIMING SIGNAL; TIMING TRACKING ACCURACY; UNILATERAL SYNCHRONIZATION SYSTEM.*

clock difference The time difference between two clocks, i.e., a measure of the separation between their respective time marks.
Note: Clock differences should be reported as algebraic quantities measured on the same time scale. The date of the measurement should be given.

Example: 1645 UT, 7 Oct. 1970:
UTC(USNO)−UTC(USAF Primary #1) = −0.9μs ± 0.2μs.

clock phase slew The changing in relative phase between a given clock signal and a phase-stable reference signal.
Note: The two signals are generally at or near the same frequency or have an integral multiple frequency relationship.

close-confinement junction *See: SINGLE HETEROJUNCTION.*

close-confinement-mesa (CCM) laser Mode-stabilized diode laser structure grown by one-step liquid-phase epitaxy above a mesa etched into the substrate. Very similar to the CDH laser except that the active layer is discontinuous on both sides of the mesa. For efficient con-

tinuous-wave operation the current is confined to the active layer by grown-in back-biased junctions on both sides of the mesa. Typical reliable output power level: 3–5 mW/facet.
Note: See for example H. Nomura, *et al.,* Tech. *Digest of the Topical Meeting on Optical Fiber Communication,* Paper THBB2, Phoenix, Arizona, April 1982.

closed waveguide A waveguide that has conducting walls, thus permitting an infinite but discrete set of propagation modes, of which relatively few are practical, each discrete mode defining the propagation constant, the field at any point being describable in terms of these modes, there being no radiation field, and discontinuities and bends causing mode conversion but not radiation. For example a metallic rectangular cross section pipe. *See also: OPEN WAVEGUIDE.*

C-message weighting A noise weighting used in a noise measuring set to measure noise on a line that would be terminated by a 500-type or similar instrument. The meter scale readings are in dBrn (C-message) or in dBrnC. *See also: MESSAGE.*

CMRR *Acronym for COMMON-MODE REJECTION RATIO.*

CNR *Acronym for CARRIER-TO-NOISE RATIO.*

CNS laser *See: CHANNEL NARROW-STRIPE DIODE LASER.*

COAT *Acronym for COHERENT OPTICAL ADAPTIVE TECHNIQUE.*

coated optics The use of optical elements or components whose optical refracting and reflecting surfaces have been coated with one or more layers of dielectric or metallic material for reducing or increasing reflection from the surfaces, either totally or for selected wavelengths and for protecting the surfaces from abrasion and corrosion. The term is usually used with reference to antireflection coatings of dielectric materials, such as magnesium fluoride, silicon monoxide, silicon oxide, titanium oxide, or zinc sulfide, for the purposes of reducing or increasing reflections and for protecting the surfaces.

coating *See: ANTIREFLECTION COATING; HIGHLY-REFLECTIVE COATING; OPTICAL FIBER COATING; OPTICAL PROTECTIVE COATING.*

co-channel interference Interference resulting from two or more transmissions in the same channel.

COD *Acronym for* CATASTROPHIC OPTICAL DAMAGE.

CODEC An assembly comprising an encoder and a decoder in the same equipment.

coded character set A set of unambiguous rules that establishes a character set and the one-to-one relationships between the characters of the set and their coded representations. *See also:* CHARACTER; CODE.

code-division multiple access (CDMA) A form of modulation whereby digital information is encoded in an expanded bandwidth format. Several transmissions can occur simultaneously within the same bandwidth with the mutual interference reduced by the degree of orthogonality of the unique codes used in each transmission. It permits a high degree of energy dispersion in the emitted bandwidth.

code-independent data communication A mode of data communication that uses a code-independent protocol and does not depend for its correct functioning on the character set or code used. *Synonym:* CODE-TRANSPARENT DATA COMMUNICATION. *See also:* DATA TRANSMISSION.

coder *Synonym:* ANALOG-TO-DIGITAL ENCODER.

code-transparent data communication *Synonym:* CODE-INDEPENDENT DATA COMMUNICATION.

coefficient *See:* ABSORPTION COEFFICIENT; ELECTROOPTIC COEFFICIENT; REFLECTION COEFFICIENT; SCATTERING COEFFICIENT; TRANSMISSION COEFFICIENT.

coherence area The area in a plane perpendicular to the direction of propagation over which light may be considered highly coherent. Commonly the coherence area is the area over which the degree of coherence exceeds 0.88. *See also:* COHERENT; DEGREE OF COHERENCE.

coherence length The propagation distance over which a light beam may be considered coherent. If the spectral linewidth of the source is $\Delta\lambda$ and the central wavelength is λ_0, the coherence length in a medium of refractive index n is approximately $\lambda_0^2/n\Delta\lambda$. *See also:* DEGREE OF COHERENCE; SPECTRAL WIDTH.

coherence time The time over which a propagating light beam may be considered coherent. It is equal to coherence length divided by the phase velocity of light in a medium; approximately given by $\lambda_0^2/c\Delta\lambda$, where λ_0 is the central wavelength, $\Delta\lambda$ is the spectral linewidth and c is the velocity of light in vacuum. *See also:* COHERENCE LENGTH; PHASE VELOCITY.

coherent Characterized by a fixed phase relationship between points on an electromagnetic wave.
Note: A truly monochromatic wave would be perfectly coherent at all points in space. In practice, however, the region of high coherence may extend only a finite distance. The area on the surface of a wavefront over which the wave may be considered coherent is called the coherence area or coherence patch; if the wave has an appreciable coherence area, it is said to be spatially coherent over that area. The distance parallel to the wave vector along which the wave may be considered coherent is called the coherence length; if the wave has an appreciable coherence length, it is said to be phase or length coherent. The coherence length divided by the velocity of light in the medium is known as the coherence time; hence a phase coherent beam may also be called time (or temporally) coherent. *See also:* COHERENCE AREA; COHERENCE LENGTH; COHERENCE TIME; DEGREE OF COHERENCE; MONOCHROMATIC.

coherent bundle *Synonym:* ALIGNED BUNDLE.

coherent light Light that has the property that at any point in time or space, particularly over an area in a plane perpendicular to the direction of propagation or over time at a particular point in space, all the parameters of the wave are predictable and are correlated. *See also:* SPACE-COHERENT LIGHT; TIME-COHERENT LIGHT.

coherent optical adaptive technique (COAT) A technique used to improve the power density of electromagnetic wavefronts, such as those of a laser beam, propagating through turbulent atmosphere, using approaches like phase conjugation, compensating phase shift aperture tagging, and image sharpening.

coherent pulse operation The method of pulse operation in which a fixed phase relationship of the carrier wave is maintained from one pulse to the next.

coherent radiation *See:* COHERENT.

collection angle The angle measured

from the (longitudinal) line between a light source and the center of a detector to a line between the light source and the circumference of the detector aperture (iris or fixed aperture between source and photodectector). By varying the iris the relative radiant flux as a function of collection angle can be determined for a diode laser or LED.

collection cone   The cone whose included apex angle is equal to twice the collection angle.

collective lens   A lens of positive power, such as a field lens, used in an optical system to refract the chief rays of image-forming bundles of rays, so that these rays will pass through subsequent optical elements of the system.
Note: If all the rays do not pass through an optical element a loss of light ensues, known as vignetting. Sometimes the term collective lens is used incorrectly to denote any lens of positive power. *See also: CONVERGING LENS.*

collimated light   A bundle of light rays in which the rays emanating from any single point in the object are parallel to one another such as the light from an infinitely distant real source, or apparent source, such as a collimator reticle. *Synonym: PARALLEL LIGHT.*

collimated transmittance   Transmittance of an optical waveguide, such as an optical fiber or integrated optical circuit, in which the light wave at the output has coherency related to the coherency at the input.

collimation   The process by which a divergent or convergent beam of radiation is converted into a beam with the minimum divergence possible for that system (ideally, a parallel bundle of rays). *See also: BEAM DIVERGENCE.*

collimator   An optical device that renders diverging or converging light rays parallel.
Note: It may be used to simulate a distant target, align the optical axes of instruments, or prepare rays for entry into the end of an optical fiber, fiber bundle, or optical thin-film.

collision   1. In a data transmission system, the situation that occurs when two or more demands are made simultaneously on equipment that can handle only one at any given instant. 2. In a computer, the situation that occurs when the same address (such as might occur when calculating hash-addresses) is ob-

tained for two different data items that are to be stored at that address. *See also: CALL COLLISION; CLEAR COLLISION.*

color   The sensation produced by light of different wavelengths in the visible spectrum.
Note: The color, shape, number of Newton's rings present when two optical surfaces are placed together, and chromatic aberration, are examples of color-related properties of light.

color-division multiplexing   (CDM)   In optical communication systems, the multiplexing of channels on a single transmission medium, such as using each color as a channel in one optical fiber or bundle of fibers.
Note: CDM is the same as frequency division multiplexing in the non-visible region of the electromagnetic frequency spectrum. Each color corresponds to a different frequency and a different wavelength.

colorimeter   An optical instrument used to compare the color of a sample with a source reference or a synthesized stimulus. For example, in a three-color colorimeter, the synthesized stimulus is produced by mixtures of three colors of fixed chromaticity, but variable luminance.

color temperature   The temperature of a blackbody that emits light of the same color as the body being considered.
Note: Color temperature is expressed in Kelvin.

coma   An aberration of a lens that causes oblique pencils of light rays from an object point or source to be imaged as a comet-shaped blur.

combinational logic element   A device having at least one output channel and zero or more input channels, all characterized by discrete states, such that at any instant the state of each output channel is completely determined by the states of the input channels at the same instant.

combiner   *See: MAXIMAL-RATIO COMBINER; SELECTIVE COMBINER.*

common-channel signaling   A signaling method using a link, common to a number of channels, necessary for the control, accounting, and management of traffic on these channels.

common-mode interference   Interference that appears between signal leads, or the terminals of a measuring circuit, and ground.

common-mode rejection ratio (CMRR) The ratio of the common-mode interference voltage at the input of a circuit, to the interference voltage at the output.

common-mode voltage 1. Any uncompensated combination of generator-receiver ground potential difference, the generator offset voltage, and the longitudinally coupled peak random noise voltage measured between the receiver circuit ground and cable with the generator ends of the cable short-circuited to ground. 2. The algebraic mean of the two voltages appearing at the receiver input terminals with respect to the receiver circuit ground.

common return A return path that is common to two or more circuits and that serves to return currents to their source or to ground.

common return offset The dc common return potential difference of a line.

communications satellite An orbiting vehicle which relays signals between communications stations. They are of two types: (a) active communications satellite—a satellite which receives, regenerates, and retransmits signals between stations; and (b) passive communications satellite—a satellite which reflects communications signals between stations.

communications sink A device which receives information, control, or other signals from communications source(s).

communications source A device which generates information, control, or other signals destined for communications sink(s).

communications subsystem A major functional part of a communications system, usually consisting of facilities and equipments essential to the operational completeness of a system.

communications system A collection of individual communications networks, transmission systems, relay stations, tributary stations, and terminal equipment capable of interconnection and interoperation to form an integral whole. *Note:* These individual components must serve a common purpose, be technically compatible, employ common procedures, respond to some form of control, and, in general, operate in unison. *See also:* COMMON CONTROL SYSTEM; ERROR-CORRECTING SYSTEM; ERROR-DETECTING SYSTEM; HYBRID SYSTEM; INTEGRATED SYSTEM; NEUTRAL DIRECT CURRENT TELEGRAPH SYSTEM; POLARENTIAL TELEGRAPH SYSTEM; PRIMARY DISTRIBUTION SYSTEM; PROTECTED WIRELINE DISTRIBUTION SYSTEM; SWITCHING SYSTEM; SYNCHRONOUS SYSTEM; SYSTEM STANDARDS; TACTICAL COMMUNICATIONS SYSTEM; WIDEBAND SYSTEM.

compaction *See:* DATA COMPACTION.

compandor A device that incorporates a compressor and an expandor, each of which may be used independently. *See also:* CLIPPER; COMPRESSOR; EXPANDOR; PEAK LIMITING.

compatibility 1. Systems for command and control, and communications are compatible with one another when necessary information can be exchanged at appropriate levels of command directly and in usable form. Communications equipments are compatible with one another if signals can be exchanged between them without the addition of buffering, translative, or similar devices for the specific purpose of achieving workable interface connections and if the equipments or systems being interconnected possess comparable performance characteristics, including suppression of undesired radiation. 2. Capability of two or more items or components of equipment or material to exist or function in the same system or environment without mutual interference. *See also:* COMMONALITY; ELECTROMAGNETIC COMPATIBILITY; INTEROPERABILITY.

compatible sideband transmission That method of independent sideband transmission wherein the carrier is deliberately reinserted at a lower level after its normal suppression to permit reception by conventional AM receivers. *Note:* The normal method of transmitting compatible SSB or AME is the carrier plus upper sideband transmission. *Synonyms:* AMPLITUDE MODULATION EQUIVALENT (AME); COMPATIBLE SSB.

compatible SSB *See:* COMPATIBLE SIDEBAND TRANSMISSION.

composited circuit A circuit which can be used simultaneously for telephony and dc telegraphy, or signaling, separation between the two being accomplished by frequency discrimination. *Synonym:* VOICE-PLUS CIRCUIT. *See also:*

SPEECH-PLUS-DUPLEX OPERATION; SPEECH PLUS SIGNALING OR TELEGRAPH.

compound lens A lens composed of two or more separate pieces of glass or other optical material. These component pieces or elements may or may not be cemented together. A common form of compound lens is a two element objective, one element being a converging lens of crown glass and the other a diverging lens of flint glass. The combination of suitable glasses or other optical materials (plastics, minerals), properly ground and polished, reduces aberrations normally present in a single lens.

compound-glass process *See: DOUBLE-CRUCIBLE PROCESS.*

compression *See: DATA COMPRESSION.*

compressor A device with a non-linear gain characteristic that acts to reduce the gain more on large input signals than it does on smaller input signals.
Note: Usually used to allow signals with a larger dynamic amplitude range to be sent through devices and circuits with a more limited range. *See also: CLIPPER; COMPANDOR; EXPANDOR; PEAK LIMITING.*

concatenation (of optical waveguides) The linking of optical waveguides, end to end.

concave Pertaining to a hollow curved surface of a given material.
Note: If a lens is imbedded in a medium, and a lens surface is concave, the contiguous surface of the medium is convex. *See also: CONVEX.*

concave lens *See: DIVERGING LENS.*

concavo-convex lens *See: MENISCUS.*

concentrator In data transmission, a functional unit that permits a common path to handle more data sources than there are channels currently available within the path.
Note: A concentrator usually provides communication capability between many low-speed, usually asynchronous channels and one or more high-speed, usually synchronous channels. Usually different speeds, codes, and protocols can be accommodated on the low-speed side. The lowspeed channels usually operate in contention and require buffering. *See also: COMPUTER; CONTENTION.*

concentric lens A lens in which the centers of curvature of the surfaces coincide.

Note: Concentric lenses thus have a constant radial thickness in all zones.

concentricity error When used in conjunction with a tolerance field to specify core/cladding geometry, the distance between the center of the two concentric circles specifying the cladding diameter and the center of the two concentric circles specifying the core diameter. *See also: CLADDING; CLADDING DIAMETER; CORE; CORE DIAMETER; TOLERANCE FIELD.*

condensing lens A lens, or system of lenses, of positive power used for condensing, i.e., converging, radiant energy from a source onto an object.

conditioned circuit A circuit that has conditioning equipment to obtain the desired characteristics for voice or data transmission.

conditioned diphase modulation A method of modulation employing both diphase modulation and signal conditioning to eliminate the dc component of a signal, to enhance timing recovery, and to facilitate transmission over VF circuits or coaxial cable facilities.

conditioned loop A loop which has conditioning equipment to obtain the desired line characteristics for voice or data transmission.

conditioned voice grade circuit A voice grade circuit that has conditioning equipment to equalize envelope or phase delay response, etc., so as to improve data transmission through the circuit. *Synonym: DATA GRADE CIRCUIT.*

conditioning equipments 1. At junctions of circuits, equipment used to match transmission levels and impedances and also to provide equalization between facilities. 2. Corrective networks used to equalize the insertion loss versus frequency characteristic and the envelope delay distortion over a desired frequency range in order to improve data transmission.

conducted interference 1. Interference resulting from radio noise or unwanted signals entering a device by direct coupling. 2. An undesired voltage or current generated within a receiver, transmitter, or associated equipment, and appearing at the antenna terminals.

conduction band A partially filled or empty energy level band in the electronic energy-band scheme characterizing a solid in which electrons can move easily in the conduction band, allowing the material to conduct electric current readily. The bandgap energy separates the conduction band from the valence band.

conductor *See: OPTICAL CONDUCTOR.*

conductor loss *See: CONNECTOR INDUCED OPTICAL CONDUCTOR LOSS.*

conduit *See: NON-COHERENT BUNDLE.*

cone *See: ACCEPTANCE CONE.*

conference operation 1. In a telephone system, that type of operation in which more than two stations can carry on a conversation. 2. In telegraph or data transmission, that form of simplex or half-duplex operation in which more than two stations may simultaneously exchange information, carry on conversations or pass messages. *Note:* In radio systems, the stations receive simultaneously, but must transmit one at a time. The common modes are "push-to-talk" (telephone) and "push-to-type" (telegraph, data transmission).

conference repeater A repeater, connecting several circuits, which receives telephone or telegraph signals from any one of the circuits and automatically retransmits them over all the others. *See also: DATA CONFERENCING REPEATER.*

confinement factor In a semiconductor laser the percentage or fraction of the optical mode power that propagates in the active layer. Symbol: Γ. *Synonyms: RADIATION CONFINEMENT FACTOR; POWER-FILLING FACTOR.*

confinement layer n- or p-type doped crystalline layer grown at or in the immediate vicinity of the active layer to provide carrier confinement and/or optical confinement in the plane perpendicular to the junction. *Note:* In DH-type devices the confinement layers sandwich the active layer, while in LOC-type devices the confinement layers sandwich the active and guide layers.

conical fiber *See: OPTICAL TAPER.*

connection *See: LASER SERVICE CONNECTION.*

connection-in-progress signal A call control signal at the DCE/DTE interface that indicates to the DTE that the establishment of the data connection is in progress and that the ready-for-data signal will follow. *See also: DATA CIRCUIT-TERMINATING EQUIPMENT.*

connections per circuit hour (CCH) A unit of traffic measurement, i.e., the number of connections established at a switching point per hour.

connector: *See: OPTICAL WAVEGUIDE CONNECTOR.*

connector-induced optical conductor loss That part of connector insertion loss usually expressed in decibels (dB) due to impurities or structural changes to the optical conductors caused by termination or handling within the connector.

connector insertion loss: *See: INSERTION LOSS.*

conservation *See: RADIANCE CONSERVATION.*

conservation of radiance: A basic principle stating that no passive optical system can increase the quantity Ln^{-2} where L is the radiance of a beam and n is the local refractive index. Formerly called "conservation of brightness" or the "brightness theorem." *See also: BRIGHTNESS; RADIANCE.*

constant *See: ABBE CONSTANT; BOLTZMANN'S CONSTANT; PROPAGATION CONSTANT.*

constricted double-heterojunction (CDH) diode laser Mode-stabilized diode laser structure grown by one-step liquid-phase epitaxy above a mesa separating two substrate channels. The active layer above the mesa assumes a convex-lens-like shape or an asymmetrical shape. Typical reliable output power levels: 3–7 mW/facet cw. *Note:* For examples see D. Botez, *IEEE J. Quantum Electron.,* vol. 17, pp. 2290–2309, Dec. 1981.

constricted double-heterojunction large-optical-cavity (CDH-LOC) diode laser Mode-stablized laser grown by one-step liquid-phase epitaxy above the mesa separating two substrate channels, and having a LOC-type structure. Both the active and guide layers vary laterally in thickness above the mesa. Reliable output power levels: 15–20 mW/facet cw and 40–50 mW/facet at 50% duty cycle. *Note:* See for example D. Botez, *Appl. Phys. Lett.,* vol. 36, pp. 190–192, Feb. 1980.

contact *See: ELECTRICAL CONTACT.*

continuously variable slope delta modulation (CVSD) A type of delta modulation in which the size of the steps of the approximated signal is progressively increased or decreased as required to make the approximated signal closely match the input analog wave. *See also:* DELTA MODULATION; MODULA-TION.

continuous operation In data transmission, a type of operation in which the master station need not stop for a reply after transmitting each message or transmission block. *See also:* DATA TRANS-MISSION.

continuous variable optical attenuator A device that attenuates the intensity of lightwaves, when inserted into an optical waveguide link, over a continuous range of dB depending upon a setting or control signal.

continuous wave (CW) A radio wave of constant amplitude and constant frequency. *See also:* INTERRUPTED CON-TINUOUS WAVE.

contrast transfer function *See:* MOD-ULATION TRANSFER FUNCTION.

convergence The bending of light rays toward each other, as by a convex or plus lens.

convergence angle The angle formed by the lines of sight of both eyes in focussing on any line, corner, surface, or part of an object. *Synonym:* CONVER-GENT ANGLE.

convergent angle *See:* CONVERGENCE ANGLE.

convergent lens *See:* CONVERGING LENS.

converging lens A lens that adds convergence to an incident bundle of light rays. One surface of a converging lens may be convexedly spherical and the other plane (plano-convex). Both may be convex (double-convex, biconvex) or one surface may be convex and the other concave (converging meniscus). *Synonyms:* CONVERGENT LENS; CONVEX LENS; COLLECTIVE LENS; CROWN LENS; POSITIVE LENS. *See also:* COLLEC-TIVE LENS.

converter *See:* ANALOG-TO-DIGITAL CONVERTER; DIGITAL-TO-ANALOG CONVERTER; DOWN-CONVERTER; PARALLEL-TO-SERIAL CONVERTER; SERIAL-TO-PARALLEL CONVERTER; SIGNAL CONVERTER; VOICE FRE-QUENCY TELEGRAPH.

convex Pertaining to a surface of an object that has its center of curvature on the same side of the surface as the material of which the object is made, thus the outside surface of a sphere or ball is convex. *See also:* CONCAVE.

convex lens *See:* CONVERGING LENS

coordinated time scale A time scale generated by electronic or mechanical devices, such as electronic clocks driven by crystal or atomic oscillators, which is coordinated by international agreement to approximate Universal Time (UT). This Coordinated Universal Time is referred to as UTC. *See also:* GREENWICH MEAN TIME; LEAP SECOND.

core The central region of an optical waveguide through which light is transmitted. *See also:* CLADDING; NORMAL-IZED FREQUENCY; OPTICAL WAVE-GUIDE.

core area The cross sectional area enclosed by the curve that connects all points nearest the axis on the periphery of the core where the refractive index of the core exceeds that of the homogeneous cladding by k times the difference between the maximum refractive index in the core and the refractive index of the homogeneous cladding, where k is a specified positive or negative constant $|k| < 1$. *See also:* CLADDING; CORE; HOMOGENEOUS CLADDING; TOLER-ANCE FIELD.

core center A point on the fiber axis. *See also:* FIBER AXIS; OPTICAL AXIS.

core diameter The diameter of the circle that circumscribes the core area. *See also:* CLADDING; CORE; CORE AREA; TOL-ERANCE FIELD.

core fiber *See:* LIQUID-CORE FIBER.

core-radii *See:* MISMATCH-OF-CORE-RADII LOSS.

corner-cube reflector *See:* TRIPLE MIR-ROR.

corner reflector *See:* TRIPLE MIRROR.

corrected lens A lens designed to be relatively free from one or more aberrations. For example, a simple lens with an aspheric surface, or a compound lens consisting of several optical elements and different glasses.

cosine emission law *Synonym:* LAM-BERT'S COSINE LAW.

cosmic noise Random noise originating outside the earth's atmosphere. *Note:* Its characteristics are similar to thermal noise. It is experienced at frequencies above about 15 MHz when highly directional antennas are pointed toward the sun or to certain other regions of the sky such as the center of the Milky Way Galaxy. *Synonym:* GALAC-TIC RADIO NOISE.

coupled modes Modes whose energies are shared. *See also: MODE.*

coupler *See: OPTICAL WAVEGUIDE COUPLER.*

coupler loss *See: SOURCE-COUPLER LOSS.*

coupler-switch-modulator *See: INTEGRATED-OPTICAL CIRCUIT FILTER-COUPLER-SWITCH-MODULATOR.*

coupling *See: MODE COUPLING.*

coupling coefficient A measure of the electrical coupling that exists between two circuits; it is equal to the ratio of the mutual impedance to the square root of the product of the self impedances of the coupled circuits, all impedances being of the same kind.

coupling efficiency The efficiency of optical power transfer between two optical components. *See also: COUPLING LOSS.*

coupling loss The power loss suffered when coupling light from one optical device to another. *See also: ANGULAR MISALIGNMENT LOSS; EXTRINSIC JOINT LOSS; GAP LOSS; INSERTION LOSS; INTRINSIC JOINT LOSS; LATERAL OFFSET LOSS.*

cpi (characters per inch) The number of characters recorded on an inch of magnetic tape.

cps (characters per second) *See: WORD.*

crank-type transverse-junction-stripe (crank-TJS) diode laser Mode-stabilized diode laser of the TJS-type, for which the Zn diffusion front has a crank-shaft-like pattern in the longitudinal direction. The devices are cleaved through areas of light propagation where Zn is not present. The device is a NAM-type device. Typical reliable power levels: 10–15 mW/facet cw.
Note: See for example Kumabe, *et al.,* *CLEO'81,* Paper FA4, Wash., DC, June 1981.

critical angle When light propagates in a homogeneous medium of relatively high refractive index (n_{high}) onto a planar interface with a homogeneous material of lower index (n_{low}), the critical angle is defined by
$$\arcsin{(n_{low}/n_{high})}.$$
Note. When the angle of incidence exceeds the critical angle, the light is totally reflected by the interface. This is termed total internal reflection. *See also: ACCEPTANCE ANGLE; ANGLE OF INCIDENCE; REFLECTION; REFRACTIVE INDEX (OF A MEDIUM); STEP INDEX PROFILE; TOTAL INTERNAL REFLECTION.*

critical frequency 1. In radio propagation by way of the ionosphere, the limiting frequency below which a wave component is reflected by, and above which it penetrates through, an ionospheric layer. 2. The limiting frequency below which a wave component is reflected by, and above which it penetrates through, an ionospheric layer at vertical incidence.
Note: The existence of the critical frequency is the result of electron limitation, i.e., the inadequacy of the existing number of free electrons to support reflection at higher frequencies.

critical radius 1. The radius of curvature of an optical fiber, containing an axially propagated electromagnetic wave, at which the field outside the fiber (that decays exponentially in a direction transverse to the direction of propagation) detaches itself from the waveguide and radiates into space because the phase front velocity must increase to maintain proper relationship with the guided wave, and this velocity cannot exceed the velocity of light, as the wave front sweeps around the curved fiber.
Note: This causes attenuation due to a radiation loss.
2. The radius of curvature of an optical fiber at which there is an appreciable propagation mode conversion loss due to the abruptness of the transition from straight to curved.
Note: For a radius of curvature greater than the critical value, the fields behave essentially as in a straight guide. For radii smaller than the critical value, considerable mode conversion takes place.

critical technical load That part of the total technical power requirement which is required for synchronous communications and automatic switching equipment.

cross connect Those connections between terminal blocks on the two sides of a distribution frame, or between terminals on a terminal block. Connections between terminals on the same block are also called straps.

cross coupling The coupling of a signal from one channel, circuit, or conductor to another, where it becomes an undesired signal.

cross modulation Intermodulation due

to the modulation of the carrier of the desired signal by the undesired signal wave.

cross talk   In an optical transmission system, leakage of optical power from one optical conductor to another.

Note: The leakage may occur by frustrated total reflection from inadequate cladding thickness or low absorptive quality. *See also: FIBER CROSSTALK.*

crosstalk   The phenomenon in which a signal transmitted on one circuit or channel of a transmission system creates an undesired effect in another circuit or channel. *See also: CROSSTALK COUPLING; FAR-END CROSSTALK; INTELLIGIBLE CROSSTALK; INTERACTION CROSSTALK; NEAR-END CROSSTALK; UNINTELLIGIBLE CROSSTALK.*

crosstalk coupling   The ratio of the power in a disturbing circuit to the induced power in the disturbed circuit observed at definite points of the circuits under specified terminal conditions, expressed in dB. *Synonym: CROSSTALK COUPLING LOSS. See also: CROSSTALK; LOSS.*

crosstalk coupling loss   *Synonym: CROSSTALK COUPLING.*

crown lens   *See: CONVERGING LENS.*

crucible process   *See: DOUBLE-CRUCIBLE PROCESS (DC).*

crystal   *See: DOUBLY-REFRACTING CRYSTAL; MULTI-REFRACTING CRYSTAL.*

crystal optics   The study of the propagation of radiant energy through crystals, especially anisotropic crystals, and their effects on polarization of electromagnetic waves, particularly light waves.

CSC   *See: CIRCUIT SWITCHING CENTER*

CSP laser   *See: CHANNEL-SUBSTRATE-PLANAR DIODE LASER.*

CSU   *See: CIRCUIT SWITCHING UNIT.*

current pumping efficiency   The fraction of current injected in the active region that is used in lasing. For oxide-stripe devices the current pumping efficiency is ≅90%, since the lateral mode confinement is determined by the injected carrier distribution. For devices where the optical mode is laterally confined by a built-in dielectric waveguide a sizable part of the current may go into spontaneous emission rather than lasing. *Note:* For examples see the CDH laser description by D. Botez, *IEEE J. Quantum Electron.,* vol. QE-17, pp. 2290–2309, Dec. 1981.

curvature   In the measurement or specification of lenses, the amount of departure from a flat surface.

Note: It is specified as the reciprocal of the radius of curvature. *See also: FIELD CURVATURE.*

curvature loss   *Synonym: MACROBEND LOSS.*

curve   *See: ABSOLUTE LUMINOSITY CURVE; LUMINOSITY CURVE.*

cutback technique   A technique for measuring fiber attenuation or distortion by performing two transmission measurements. One is at the output end of the full length of the fiber. The other is within 1 to 3 meters of the input end, access being had by "cutting back" the test fiber. *See also: ATTENUATION.*

cutoff wavelength   That wavelength greater than which a particular waveguide mode ceases to be a bound mode.

Note. In a single mode waveguide, concern is with the cutoff wavelength of the second order mode. *See also: MODE.*

CVD   *Acronym for CHEMICAL VAPOR DEPOSITION.*

CVPO   *Acronym for CHEMICAL VAPOR-PHASE OXIDATION PROCESS.*

CVSD   *Acronym for CONTINUOUSLY VARIABLE SLOPE DELTA MODULATION.*

CW diode laser   Diode laser that can be operated in the continuous-wave (CW) mode; that is, while being driven by direct current. For optical communications, diode lasers capable of sustained operation at room temperature are absolutely necessary.

cyclic distortion   In telegraphy, distortion which is neither characteristic, bias, nor fortuitous, and which, in general, has a periodic character.

Note: Its causes are, for example, irregularities in the duration of contact time of the brushes of a transmitter distributor or interference by disturbing alternating currents.

cylindrical lens   A lens with a cylindrical surface.

Note: Cylindrical lenses are used in rangefinders to introduce astigmatism in order that a point-like source may be imaged as a line of light. By combining cylindrical and spherical surfaces, an optical system can be designed that gives a certain magnification in a given azimuth of the image and a different magnification at right angles in the same image plane. Such a system is designated as being anamorphic.

D* (pronounced "D-star") A figure of merit often used to characterize detector performance, defined as the reciprocal of noise equivalent power (NEP), normalized to unit area and unit bandwidth.

$$D^* = \sqrt{A(\Delta f)}/NEP,$$

where A is the area of the photosensitive region of the detector and (Δf) is the effective noise bandwidth. *Synonym: SPECIFIC DETECTIVITY. See also: DETECTIVITY; NOISE EQUIVALENT POWER.*

D-A *Acronym for DIGITAL-TO-ANALOG CONVERTER.*

damping 1. The progressive diminution with time of certain quantities characteristic of a phenomenon. 2. The progressive decay with time in the amplitude of the free oscillations in a circuit.

dark adaptation The ability of the human eye to adjust itself to low levels of illumination.

dark current The external current that, under specified biasing conditions, flows in a photosensitive detector when there is no incident radiation.

data Any representations such as characters or analog quantities to which meaning is or might be assigned. *See also: ANALOG DATA; BUFFER; DATA COMMUNICATION; DATA TRANSMISSION; DIGITAL DATA; SIGNALING TIME SLOT.*

data bus In an optical communication system, an optical waveguide used as a common trunk line to which a number of terminals can be interconnected using optical couplers.

data bus coupler In an optical communication system, a component that interconnects a number of optical waveguides and provides an inherently bidirectional system by mixing and splitting all signals within the component.

data circuit connection The interconnection of a number of links or trunks, on a tandem basis, by means of switching equipment to enable data transmission to take place among data terminal equipment. *See also: CIRCUIT.*

data circuit-terminating equipment (DCE) The interfacing equipment sometimes required to couple the DTE (data terminal equipment) into a transmission circuit or channel and from a transmission circuit or channel into the DTE. *Synonyms: DATA COMMUNICATIONS EQUIPMENT (Deprecated); DATA SET (Deprecated). See also: CONNECTION-IN-PROGRESS SIGNAL; DATA MODE; DATA SINK; DATA SOURCE; DATA TERMINAL EQUIPMENT (DTE).*

data communication Data transfer between data source and data sink via one or more data links according to a protocol. *See also: DATA; DATA TRANSMISSION; MASTER STATION; PRIMARY STATION; SECONDARY STATION.*

data communication control procedure A means used to control the orderly communication of information among stations in a data communication network.

data communications equipment Deprecated. *See: DATA CIRCUIT-TERMINATING EQUIPMENT (DCE).*

data compaction Pertaining to the reduction of space, bandwidth, cost, and time for the generation, transmission, and storage of data by employing techniques designed to eliminate repetition, remove irrelevancy, and employ special coding.
Note: Some data compaction methods employ fixed tolerance bands, variable tolerance bands, slope-keypoints, sample changes, curve patterns, curve fitting, floating-point coding, variable precision coding, frequency analysis, and probability analysis. (Simply squeezing non-

compacted data into a smaller space, for example, by transferring data on punched cards onto magnetic tape, is not considered data compaction.) *See also: DATA COMPRESSION.*

data compression 1. A method of increasing the amount of data that can be stored in a given space or contained in a given message length. 2. A method of reducing the amount of storage space required to store a given amount of data or reducing the length of message required to transfer a given amount of information. *See also: DATA COMPACTION.*

data concentrator *See: CONCENTRATOR.*

data conferencing repeater A device that enables a group of users to operate such that if any one user transmits a message it will be received by all others in the group. *Synonym: TECHNICAL CONTROL HUBBING REPEATER. See also: CONFERENCE REPEATER; NETWORK.*

data grade circuit *Synonym: CONDITIONED VOICE GRADE CIRCUIT.*

data link A communications link suitable for transmission of data. *See also: LINK; OPTICAL DATA LINK; TACTICAL DIGITAL INFORMATION LINK.*

data mode The state of a DCE when connected to a communication channel but not in a talk or dial mode. *See also: DATA CIRCUIT-TERMINATING EQUIPMENT.*

data phase A phase of a data call during which data signals may be transferred between DTEs that are interconnected via the network. *See also: DATA TERMINAL EQUIPMENT; DATA TRANSMISSION.*

data set Deprecated. *See: DATA CIRCUIT-TERMINATING EQUIPMENT.*

data signaling rate A measure of signaling speed given by:

$$DSR = \sum_{i=1}^{m} (1/T_i) \log_2 n_i$$

where DSR is the data signaling rate, m is the number of parallel channels, T_i is the minimum interval for the i-th channel expressed in seconds, n_i is the number of significant conditions of the modulation in the i-th channel. Data signaling rate is expressed in bits per second (b/s).

Note: 1. For a single channel (serial transmission) it reduces to $(1/T)\log_2 n$; with a two-condition modulation $(n=2)$, it is $1/T$.

Note: 2. For a parallel transmission with equal minimum intervals and equal number of significant conditions on each channel, it is $(m/T) \log_2 n$; in case of a two-condition modulation, this reduces to (m/T).

Note: 3. In synchronous binary signaling, the data signaling rate in bits per second is numerically the same as the modulation rate expressed in baud. Signal processors, such as four-phase modems, cannot change the data signaling rate, but the modulation rate depends on the line modulation scheme, according to note 2. For example, in a 2400 b/s 4-phase sending modem, the signaling rate is 2400 b/s on the serial input side, but the modulation rate is only 1200 baud on the 4-phase output side. *See also: BITS PER SECOND; DATA TRANSFER RATE; SIGNALING.*

data signaling-rate transparency A network characteristic that enables the transfer of data between one user and another at data signaling rates that may vary within certain limits. *See also: NETWORK; SIGNALING; TRANSPARENCY.*

data sink A device which receives data signals from data source(s). *See also: DATA CIRCUIT-TERMINATING EQUIPMENT.*

data source A device which generates data signals destined for data sink(s). *See also: DATA CIRCUIT-TERMINATING EQUIPMENT (DCE).*

data stream A sequence of binary digits used to represent information for transmission.

data subscriber terminal equipment (DSTE) A general purpose AUTODIN terminal device consisting of all necessary equipment: (a) to provide AUTODIN interface functions; (b) to perform code conversions; and (c) to transform punched card messages, punched paper tape, or magnetic tape to electrical signals for transmission, and the reverse of this process.

data switching exchange (DSE) Equipment installed at a single location to switch data traffic.

Note: A data switching exchange may provide only circuit switching, only

packet switching, or both. *See also:*
AUTOMATIC EXCHANGE; SWITCH-
ING CENTER; SYNCHRONOUS DATA
NETWORK.

data terminal equipment (DTE) 1.
Equipment consisting of digital end instruments that convert the user information into data signals for transmission, or
reconvert the received data signals into
user information. 2. The functional unit
of a data station that serves as a data
source or a data sink and provides for
the data communication control function
to be performed in accordance with link
protocol.
Note: The DTE may consist of a single
piece of equipment which provides all
the required functions necessary to permit the user to intercommunicate, or it
may be an interconnected subsystem of
multiple pieces of equipment, including
communications security equipment, to
perform all the required functions. *See
also: AUTOMATIC ANSWERING;
AUTOMATIC CALLING; AUTOMATIC
SEQUENTIAL CONNECTION; BIT SYN-
CHRONOUS OPERATION; CALL AC-
CEPTED SIGNAL; CALL COLLISION;
CALL-NOT-ACCEPTED SIGNAL; CALL
PROGRESS SIGNAL; CALL SET-UP
TIME; CALLED-LINE IDENTIFICATION
SIGNAL; CALLING-LINE IDENTIFICA-
TION SIGNAL; CALLS BARRED FACILI-
TY; CLEAR COLLISION; CLEAR CON-
FIRMATION SIGNAL; DATA CIR-
CUIT-TERMINATING EQUIPMENT;
DATA PHASE; DTE CLEAR SIGNAL;
DTE WAITING SIGNAL; FLOW CON-
TROL PROCEDURE; INTERCHANGE
CIRCUIT; NETWORK CONTROL
PHASE; REQUEST DATA TRANSFER;
VIRTUAL CALL CAPABILITY.*

data transfer rate The number of bits,
characters, or blocks per unit time passing between corresponding equipments
in a data transmission system. It is expressed in terms of bits, characters, or
blocks per second, minute, or hour. *See
also: DATA SIGNALING RATE.*

data transfer time The time that
elapses between the initial offering of a
unit of user data to a network by transmitting data terminal equipment and the
complete delivery of that unit to receiving data terminal equipment.

data transmission The sending of data
from one place to another by means of
signals over a channel. *See also: ADDRESS*

*FIELD; BACKWARD CHANNEL; BASIC
STATUS; CODE-INDEPENDENT DATA
COMMUNICATION; COLLISION;
COMMAND; COMMAND FRAME;
CONTINUOUS OPERATION; DATA;
DATA COMMUNICATION; DATA
COMMUNICATION CONTROL CHAR-
ACTER; DATA LINK ESCAPE CHAR-
ACTER; DATA PHASE; EXCEPTION
CONDITION; FLAG SEQUENCE; FOR-
WARD CHANNEL; FRAME; HIGH-LEV-
EL DATA LINK CONTROL; INFOR-
MATION FIELD; INTERACTIVE DATA
TRANSACTION; LINK LEVEL; MASTER
STATION; NONSYNCHRONOUS
DATA TRANSMISSION CHANNEL;
PACKET-SWITCHED DATA TRANS-
MISSION SERVICE; PUBLIC DATA NET-
WORK; PUBLIC DATA TRANSMIS-
SION SERVICE; SYNCHRONOUS
DATA-LINK CONTROL; SYNCHRO-
NOUS DATA NETWORK; SYNCHRO-
NOUS NETWORK; TRANSMISSION;
TRANSMIT FLOW CONTROL; UN-
NUMBERED COMMAND; UNNUM-
BERED RESPONSE; UNSUCCESSFUL
CALL.*

data transmission circuit The transmission media and intervening equipment involved in the transfer of data
between DTEs.
Note: A data transmission circuit includes
the signal conversion equipment. A data
transmission circuit may support the
transfer of information in one direction
only, in either direction alternately, or in
both directions simultaneously. (NCS)
See also: CHANNEL.

dating format The time of an event on
the Universal Time System given in the
following sequence: Hours, Day, Month,
Year; e.g., 1445 UT, 23 August 1971.
Note: The hour is designated for a 24-
hour system.

dB *Symbol for DECIBEL.* The standard unit
for expressing transmission gain or loss
and relative power ratios. The decibel is
one tenth the size of a Bel, which is too
large a unit for convenient use. Both
units are expressed in terms of the logarithm to the base 10 of a power ratio, the
decibel formula being:

$$dB = 10 \log_{10}(P_1/P_2)$$

Power ratios may be expressed in terms
of voltage or current. If the impedances
for both the power measurements are
the same, they cancel out in the power

ratio so the formulas in terms of voltage or current become as follows:

$$dB = 10 \log_{10}[(E_1{}^2/R_1)/(E_2{}^2/R_2)]$$
$$= 10 \log_{10}[(I_1{}^2R_1)/(I_2{}^2 R_2)]$$

If $R_1 = R_2$, then:

$$dB = 10 \log_{10}(E_1{}^2/E_2{}^2)$$
$$= 10 \log_{10}(I_1{}^2/I_2{}^2)$$
$$= 20 \log_{10}(E_1/E_2)$$
$$= 20 \log_{10}(I_1/I_2)$$

dBa, dBrn adjusted   Weighted noise power, in dB referred to 3.16 picowatts (-85 dBm), which is 0 dBa. Use of F1A-line or HA1-receiver weighting shall be indicated in parentheses as required.
Note: A one milliwatt, 1000 Hz tone will read $+85$ dBa, but the same power as white noise, randomly distributed over a 3 kHz band (nominally 300 to 3300 Hz), will read $+82$ dBa, due to the frequency weighting. *See also: NOISE WEIGHTING.*

dBa(F1A)   Weighted noise power in dBa, measured by a noise measuring set with F1A-line weighting.
Note: F1A weighting is obsolete for DoD applications.

dBa(HA1)   Weighted noise power in dBa, measured across the receiver of a Western Electric 302-type or similar subset, by a noise measuring set with HA1-receiver weighting.
Note: HA1 weighting is obsolete for DoD applications.

dBa0   Noise power in dBa referred to or measured at a zero transmission level point (0TLP), also called a point of zero relative transmission level (0 dBr).
Note: It is preferred to convert noise readings from dBa to dBa0, as this makes it unnecessary to know or state the relative transmission level at point of actual measurement.

dBm   dB referred to one milliwatt; employed in communication work as a measure of absolute power values. Zero dBm equals one milliwatt.
Note: In DoD practice unweighted measurement is normally understood, applicable to a certain bandwidth which must be stated or implied. In European practice, psophometric weighting may be implied, as indicated by context; equivalent to dBm0p, which is preferred. *See also: NEPER.*

dBm0   Noise power in dBm referred to or measured at a zero transmission level point (0TLP). The 0TLP is also called a point of zero relative transmission level (0 dBr0).
Note: Some international documents use dBm0 to mean noise power in dBm0p (psophometrically weighted dBm0). In DoD practice, dBm0 is not so used.

dBm0p   Noise power in dBm0, measured by a psophometer or noise measuring set having psophometric weighting. *See also: NOISE WEIGHTING.*

dBm(psoph)   A unit of noise power in dBm, measured with psophometric weighting. For conversion to other weighted units:

$$dBm(psoph) = [10 \log_{10} pWp] - 90$$
$$= dBa - 84$$

dBr   The power difference expressed in dB between any point and a reference point selected as the zero relative transmission level point.
Note: Any power expressed in dBr does not specify the absolute power. It is a relative measurement only. *See also: PSOPHOMETRIC WEIGHTING; TRANSMISSION LEVEL; TRANSMISSION LEVEL POINT.*

DBR laser   *See: DISTRIBUTED BRAGG-REFLECTOR LASER.*

dBrn   (Decibels Above Reference Noise). Weighted noise power in dB referred to 1.0 picowatt. Thus, 0dBrn $= -90$ dBm. Use of 144-line, 144-receiver or C-message weighting, or flat weighting, shall be indicated in parentheses as required.
Note: 1. With C-message weighting, a one milliwatt, 1000 Hz tone will read $+90$ dBrn, but the same power as white noise, randomly distributed over a 3 kHz band will read approximately $+88.5$ dBrn (rounded off to $+88$ dBrn), due to the frequency weighting.
Note: 2. With 144 weightings, a 1 mW, 1000 Hz tone will also read $+90$ dBrn, but the same 3 kHz white noise power will read only $+82$ dBrn, due to the different frequency weighting. *See also: NOISE WEIGHTING.*

dBrn (144 line)   Weighted noise power in dBrn, measured by a noise measuring set with 144-line weighting.

dBrnC   Weighted noise power in dBrn, measured by a noise measuring set with C-message weighting. *See also: CIRCUIT NOISE LEVEL.*

dBrnC0   Noise power in dBrnC referred

to or measured at a zero transmission level point (0TLP).

dBrn(f₁-f₂) Flat noise power in dBrn, measured over the frequency band between frequencies f_1 and f_2. *See also: FLAT WEIGHTING; NOISE WEIGHTING.*

dBW Decibels referred to one watt.

DC *Acronym for DOUBLE CRUCIBLE PROCESS.*

DCE *Acronym for DATA CIRCUIT-TERMINATING EQUIPMENT.*

DCPSK *Acronym for DIFFERENTIALLY COHERENT PHASE-SHIFT KEYING.*

decibel *See: dB.*

decode 1. To convert data by reversing the effect of some previous encoding. 2. To interpret a code. 3. To convert encoded text into its equivalent plain text by means of a code. (This does not include solution by cryptanalysis.) *See also: CODE; ENCODE.*

decoding *See: ANALOG DECODING.*

decollimation In a light-wave guide, such as an optical fiber or integrated optical circuit, the spreading or divergence of light due to internal and end effects, such as curvature, irregularities of surfaces, erratic variations in refractive indices, occlusions, and other blemishes that may cause dispersion, absorption, scattering, deflection, diffraction, reflection, refraction, or other effects.

deemphasis In frequency modulation, a process of reducing the amplitude of the high frequencies after their detection to restore the frequency components to their original relative level. *See also: PRE-EMPHASIS NETWORK.*

deep-Zn diffused laser Mode-stabilized laser for which Zn is diffused through a mask on the diode p-side such that the Zn front penetrates in the active layer. The diffusion creates lateral mode control as well as tight current confinement.
Note: For example see Ueno, *et al., IEEE J. Quantum Electron.,* vol. 15, pp. 1189–1195, 1979.

defect absorption *See: ATOMIC DEFECT ABSORPTION.*

degradation *See: CATASTROPHIC DEGRADATION; GRADUAL DEGRADATION.*

degree of coherence A measure of the coherence of a light source; the magnitude of the degree of coherence is equal to the visibility, V, of the fringes of a two-beam interference experiment, where

$$V = \frac{I_{max} - I_{min}}{I_{max} + I_{min}}$$

I_{max} is the intensity at a maximum of the interference pattern, and I_{min} is the intensity at a minimum.
Note. Light is considered highly coherent when the degree of coherence exceeds 0.88, partially coherent for values less than 0.88, and incoherent for "very small" values. *See also: COHERENCE AREA; COHERENCE LENGTH; COHERENT; INTERFERENCE.*

degree of individual distortion of a particular significant instant As applied to a modulation or a demodulation, the ratio to the unit interval of the maximum displacement, expressed algebraically, of this significant instant from an ideal instant. This displacement is considered positive when a significant instant occurs after the ideal instant. The degree of individual distortion is usually expressed as a percentage. *See also: DISTORTION.*

degree of isochronous distortion The ratio to the unit interval of the maximum measured difference, irrespective of sign, between the actual and the theoretical intervals separating any two significant instants of modulation (or demodulation), these instants being not necessarily consecutive (usually expressed as a percentage).
Note: The result of the measurement should be completed by an indication of the period, usually limited, of the observation. For a prolonged modulation (or demodulation) it will be appropriate to consider the probability that an assigned value of the degree of distortion will be exceeded.

degree of start-stop distortion 1. The ratio to the unit interval of the maximum measured difference, irrespective of sign, between the actual and theoretical intervals separating any significant instant of modulation (or of demodulation) from the significant instant of the start element immediately preceding it. 2. The highest absolute value of individual distortion affecting the significant instants of a start-stop modulation.
Note: The degree of distortion of a start-stop modulation (or demodulation) is usually expressed as a percentage. Dis-

tinction can be made between the degree of late (or positive) distortion and the degree of early (or negative) distortion.

delay *See: ABSOLUTE DELAY; BLOCK TRANSFER TIME; MAXIMUM BLOCK TRANSFER TIME; RECEIVE-AFTER-TRANSMIT TIME DELAY; RECEIVER ATTACK-TIME DELAY; RECEIVER RELEASE-TIME DELAY; ROUND-TRIP DELAY TIME; TRANSMIT-AFTER-RECEIVE TIME DELAY; TRANSMITTER ATTACK-TIME DELAY; TRANSMITTER RELEASE-TIME DELAY.*

delay distortion The distortion of a wave form made up of two or more different frequencies, caused by the difference in arrival time of each frequency at the output of a transmission system. *Synonyms: PHASE DISTORTION; TIME-DELAY DISTORTION. See also: ABSOLUTE DELAY; WAVEGUIDE DELAY DISTORTION.*

delay equalizer A corrective network which is designed to make the phase delay or envelope delay of a circuit or system substantially constant over a desired frequency range. *See also: ABSOLUTE DELAY.*

delay spread *See: MULTIMODE GROUP-DELAY SPREAD.*

delta modulation A technique for converting an analog signal to a digital signal. The technique approximates the analog signal with a series of segments. The approximated signal is compared to the original analog wave to determine an increase or decrease in relative amplitude. The decision process for establishing the state of successive binary digits is determined by this comparison. Only the change of information, an increase or decrease of the signal amplitude from the previous sample, is sent; thus, a no change condition remains at the same 0 or 1 state of the previous sample. There are several variations to the simple delta modulation system. *See also: CONTINUOUSLY VARIABLE SLOPE DELTA MODULATION.*

demand assignment An operational technique whereby various users share a satellite capacity on a real-time demand basis. That is, a user needing to communicate with another user of the network activates the required circuit. Upon completion of the call, the circuit is deactivated and the capacity is available for other users. This service is analogous in many ways to an ordinary telephone switching network that provides common trunking for many subscribers through a limited size trunk group on a demand basis.

democratically synchronized network A mutually synchronized network in which all clocks in the network are of equal status and exert equal amounts of control on the others, the network operating frequency being the mean of the natural (uncontrolled) frequencies of the population of clocks. *See also: DESPOTICALLY SYNCHRONIZED NETWORK; FREQUENCY AVERAGING; OLIGARCHICALLY SYNCHRONIZED NETWORK.*

demodulation The process wherein a wave resulting from previous modulation is processed to derive a wave having substantially the characteristics of the original modulating wave. *See also: RESTITUTION.*

demultiplex (DEMUX) The inverse of multiplex. *See also: MULTIPLEX.*

DEMUX *Acronym for DEMULTIPLEX.*

departure angle The angle between the axis of the main lobe of an antenna pattern and the horizontal plane at the transmitting antenna. *Synonym: TAKEOFF ANGLE.*

deposition process *See: MODIFIED CHEMICAL VAPOR DEPOSITION PROCESS; PLASMA-ACTIVATED CHEMICAL VAPOR DEPOSITION PROCESS (PACVD).*

design margin *Synonym: rf POWER MARGIN.*

despotically synchronized network A synchronized network in which a unique master clock exists with full power to control all other clocks. *See also: DEMOCRATICALLY SYNCHRONIZED NETWORK; OLIGARCHICALLY SYNCHRONIZED NETWORK.*

destuffing The controlled deletion of digits from a stuffed digital signal to recover the original signal prior to stuffing.
Note: The deleted information is transmitted via a separate low capacity time slot. Synonyms: NEGATIVE JUSTIFICATION; NEGATIVE PULSE STUFFING. See also: BIT STUFFING; NOMINAL BIT STUFFING RATE.

detectivity The reciprocal of noise equivalent power (NEP). *See also: NOISE EQUIVALENT POWER (NEP).*

detector A device responsive to the presence of a stimulus. *See also: EXTERNAL*

PHOTOEFFECT DETECTOR; INTERNAL PHOTOEFFECT DETECTOR; OPTICAL DETECTOR; PHOTODETECTOR; PHOTON DETECTOR.

detector coupling *See: FIBER-DETECTOR COUPLING.*

detector noise-limited operation In optical communication system operations, the situation that occurs when the amplitude of pulses, rather than their width, limits the distance between repeaters.

Note: In this regime of operation, the losses are sufficient to attenuate the amplitude of the pulse so much, in relation to the detector noise level, to prevent an intelligent decision based on the presence or absence of a pulse in the intelligent signal. *See also: DISPERSION-LIMITED OPERATION.*

deviation angle 1. The angular change in direction of a light ray after crossing the interface between two different media. 2. The angle through which a ray of light is bent by reflection or refraction.

deviation ratio In a frequency modulation system, the ratio of the maximum frequency deviation to the maximum modulating frequency of the system under specified conditions.

device *See: OPTOELECTRONIC DEVICE; PHOTOCONDUCTIVE DEVICE.*

devitrification The changing of glass from the vitreous (glassy) state to a crystalline state, thus greatly changing most of its optical properties, usually for the worse for optical purposes, such as reduced light transmission in optical fibers and integrated optical circuits.

DFB laser *See: DISTRIBUTED FEEDBACK LASER.*

DFSK *Acronym for DOUBLE FREQUENCY-SHIFT KEYING.*

DH laser *See: DOUBLE-HETEROJUNCTION DIODE*

diad *Synonym: DIBIT.*

diameter *See: BEAM DIAMETER; FIBER CORE DIAMETER; FIBER DIAMETER.*

dibit A group of two bits. The four possible states for a dibit are 00, 01, 10, and 11. *Synonym: DIAD.*

dichroic Pertaining to the quality of dichroism.

dichroic filter An optical filter designed to transmit light selectively according to wavelength (most often, a high-pass or low-pass filter). *See also: OPTICAL FILTER.*

dichroic mirror A mirror designed to reflect light selectively according to wavelength. *See also: DICHROIC FILTER.*

dichroism In anisotropic materials, such as some crystals, the selective absorption of light rays vibrating in one particular plane relative to the crystalline axes, but not those vibrating in a plane at right angles thereto.

Note: As applied to isotropic materials, this term refers to the selective reflection and transmission of light as a function of wavelength regardless of its plane of vibration. The color of such materials, as seen by transmitted light, varies with the thickness of material examined. *Synonyms: DICHROMATISM; POLYCHROMATISM.*

dichromatism *See: DICHROISM.*

die A crystalline piece of material such as a laser chip or submounts (Si, BeO, diamond) used in the assembly of optoelectronic devices.

dielectric constant The relative permittivity of a medium with respect to that of the vacuum. The dielectric constant is generally a complex quantity. The square root of the dielectric constant has real and imaginary parts, which by definition correspond to the index of refraction and the extinction coefficient, respectively. *See: INDEX-OF-REFRACTION; EXTINCTION COEFFICIENT. Symbols:* κ; ϵ

Note: See for example H. Kressel and J. K. Butler, *Semiconductor Lasers and Heterojunction LEDs*, Academic Press, NY, 1977.

dielectric film *See: MULTILAYER DIELECTRIC FILM.*

dielectric filter *See: INTERFERENCE FILTER.*

dielectric optical waveguide *See: SLAB-DIELECTRIC OPTICAL WAVEGUIDE.*

differentially coherent phase-shift keying (DCPSK) A method of modulation in which information is encoded in terms of phase changes, rather than absolute phases, and detected by comparing phases of adjacent bits.

Note: The carrier pulses used are of constant amplitude, angular frequency, and duration, but of different relative phase. In detection a phase comparison is made of successive samples, and information is conveyed by the phase transitions between carrier and pulses rather than by the absolute phases of the pulses.

differential mode attenuation The variation in attenuation among the propagating modes of an optical fiber.

differential mode delay The variation in propagation delay that occurs because of the different group velocities of the modes of an optical fiber. *Synonym: MULTIMODE GROUP DELAY. See also: GROUP VELOCITY; MODE; MULTIMODE DISTORTION.*

differential-mode interference 1. Interference causing a change in potential of one side of a signal transmission path relative to the other side. 2. Interference resulting from an interference current path coinciding with the signal path.

differential modulation A type of modulation in which the choice of the significant condition for any signal element is dependent on the choice for the previous signal element. Delta modulation is an example.

differential phase-shift keying (DPSK) A method of modulation employed for digital transmission. In DPSK, each signal element is a change in the phase of the carrier with respect to its previous phase angle.
Note: 1. In DPSK systems designed so that the carrier can assume only two different phase angles, then each change of phase (signal element) carries one bit of information, i.e., the bit rate equals the modulation rate.
Note: 2. If the number of recognizable phase angles is increased to 4, then 2 bits of information can be encoded into each signal element. Likewise, 8 phase angles can encode 3 bits in each signal element, i.e., unit interval, etc.

differential quantum efficiency In an optical source or detector, the slope of the curve relating output quanta to input quanta.

diffraction The deviation of a wavefront from the path predicted by geometric optics when a wavefront is restricted by an opening or an edge of an object.
Note: Diffraction is usually most noticeable for openings of the order of a wavelength. However, diffraction may still be important for apertures many orders of magnitude larger than the wavelength. *See also: FAR-FIELD DIFFRACTION PATTERN; NEAR-FIELD DIFFRACTION PATTERN.*

diffraction grating An array of fine, parallel, equally spaced reflecting or transmitting lines that mutually enhance the effects of diffraction to concentrate the diffracted light in a few directions determined by the spacing of the lines and the

wavelength of the light. *See also: DIFFRACTION.*

diffraction grating spectral order The integers that distinguish the different directions of each member of a family of light rays emerging from a diffraction grating. For example when a beam of parallel rays of monochromatic light pass through a diffraction grating. The emergent rays that have remained undeviated belong to the zero spectral order, but the light flux in the family of deviated rays that emerge after diffraction at the grating exhibit pronounced maxima along well defined and enumerable directions, on each side of the undeviated beams.
Note: The integers that are assigned to distinguish these directions mark the spectral orders. *See also: GRATING CHROMATIC RESOLVING POWER.*

diffraction limited A beam of light is diffraction limited if: a) the farfield beam divergence is equal to that predicted by diffraction theory, or b) in focusing optics, the impulse response or resolution limit is equal to that predicted by diffraction theory. *See also: BEAM DIVERGENCE ANGLE; DIFFRACTION.*

diffraction region The region beyond the radio horizon.

diffuse density The logarithm to the base 10 of the reciprocal of diffuse transmittance.

diffused optical waveguide An optical-wavelength electromagnetic waveguide consisting of a substrate, such as an optical-quality single crystal of zinc selenide (ZnSe) or cadmium sulphide (CdS), into the outer layers of which a diffusant, such as cadmium to replace the selenium in ZnSe, or selenium to replace the sulphur, has been diffused to a depth of a few microns, thus producing a lower index of refraction on the outside, thus producing a waveguide with a graded index of refraction. *See also: STRIP-LOADED DIFFUSED OPTICAL WAVEGUIDE.*

diffuse reflectance 1. The ratio of light flux reflected diffusely in all directions to the total flux at incidence, specular reflection excluded. 2. The reflectance of a sample relative to a perfectly diffusing, and perfectly reflecting standard with 45-degree incidence angle and observation along the perpendicular to the surface. *Synonym: TOTAL DIFFUSE REFLECTANCE.*

diffuse reflection *See: REFLECTION.*

diffuse transmittance 1. The transmittance measured with diffusely incident flux. 2. The ratio of the flux diffusely transmitted in all directions to the total incident flux.

diffusion The scattering of light by reflection or transmission.

Note: Diffuse reflection results when light strikes an irregular surface such as a frosted window or the surface of a frosted or coated light bulb. When light is diffused, no definite image is formed.

diffusion current Current set by a gradient in carrier concentration. Carrier leakage out of the active region in a diode laser is mostly due to diffusion.

diffusion length The mean distance a carrier travels in bulk semiconductor material before it recombines via radiative or non-radiative processes. For a given optoelectronic material type, the longer the diffusion length the higher the quality of the material is.

digit A symbol, numeral, or graphic character that represents an integer, e.g., one of the decimal characters 0 to 9, or one of the binary characters 0 or 1.

Note: In a given numeration system the number of allowable different digits, including zero, is always equal to the radix. (NCS) *See also: BIT; MAXIMUM STUFFING RATE; SYMMETRICAL BINARY CODE; SYNCHRONIZATION BIT.*

digital alphabet A coded character set in which the characters of an alphabet have a one-to-one relationship with their coded representations. *See also: CODE CHARACTER; CODE SET; DIGITAL COMBINING; DIGITIZE.*

digital block A set of multiplexed equipment that includes one or more data channels and associated circuitry.

Note: Digital blocks are usually designated in terms of signaling speed, for example, a 134.5 baud digital block. This digital block is contrasted with a block of data. *See also: MULTIPLEX.*

digital combining A method of interlacing digital data signals, in either synchronous or asynchronous mode, without converting the data into a quasi-analog signal. *See also: DIGITAL ALPHABET; DIVERSITY COMBINER.*

digital data 1. Data represented by discrete values or conditions, as opposed to analog data. 2. A discrete representation of a quantized value of a variable, i.e., the representation of a number by digits, perhaps with special characters and the "space" character.

digital error A single-digit inconsistency between the signal actually received and the signal that should have been received. *See also: CHARACTER-COUNT AND BIT-COUNT INTEGRITY; ERROR.*

digital frequency modulation The transmission of digital data by frequency modulation of a carrier, as in binary FSK.

digital modulation The process of varying one or more parameters of a carrier wave as a function of two or more finite and discrete states of a signal.

digital signal 1. A nominally discontinuous electrical signal that changes from one state to another in discrete steps. 2. A signal that is timewise discontinuous, i.e., discrete, and can assume a limited set of values.

Note: 1. The electrical signal could be changed in its amplitude or polarity.

Note: 2. Analog signals may be converted to digital signals by sampling and quantizing.

digital switch Switching equipment designed, designated, or used to connect circuits between users for transmission of digital signals.

digital switching A process in which connections are based on digital signals without converting them to analog signals.

digital synchronization *See: BIT SYNCHRONOUS OPERATION.*

digital-to-analog (D-A) converter A device that converts a digital input signal to an analog output signal carrying equivalent information. *See also: ANALOG-TO-DIGITAL (A-D) CONVERTER.*

digital transmission group A number of voice channels or a number of data channels or both that are combined into a digital bit stream for transmission over various communications media. *See also: TRANSMISSION.*

digital voice transmission Transmission of analog voice signals that have been converted into digital signals; for example, pulse code modulation (PCM) of analog voice signals.

digitize To convert an analog signal to a digital signal. *See also: DIGITAL ALPHABET.*

digit position The position in time or space into which a representation of a digit may be placed.

digit time slot In a bit stream, the time interval allocated to a single digit.

diode *See: DOUBLE HETEROJUNCTION DIODE; FIVE-LAYER FOUR HETERO-*

JUNCTION DIODE; MONORAIL DOUBLE-HETEROJUNCTION DIODE; INJECTION LASER DIODE; LASER DIODE; LIGHT-EMITTING DIODE (LED); LARGE OPTICAL-CAVITY DIODE; PIN DIODE; RESTRICTED EDGE-EMITTING DIODE (REED); SUPERLUMINESCENT DIODE (SLD).

diode coupler *See: AVALANCHE PHOTODIODE COUPLER; LASER DIODE COUPLER; LIGHT-EMITTING DIODE COUPLER; POSITIVE-INTRINSIC-NEGATIVE PHOTODIODE COUPLER.*

diode laser *Synonym: INJECTION LASER DIODE (ILD).*

diopt *See: DIOPTER.*

diopter A unit of refractive power of a lens or prism, equal to the reciprocal of the focal length in meters.

diplex operation Simultaneous one-way transmission or reception of two independent signals using a common element, such as a single antenna or channel, e.g., operation of two or more radio transmitters on different frequencies using one antenna. *See also: DUPLEX OPERATION.*

direct bandgap semiconductors Semiconductor compounds for which photons are emitted primarily by electron-hole recombination without the need for lattice vibrations (i.e., phonons). The corresponding optical transitions are called direct transitions. Only direct-bandgap semiconductors can be used for lasing material. *Examples:* GaAs; InP; $A1_xGa_{1-x}As$ for $0 < x < 0.4$.

direct coupling In optical waveguides, such as optical fibers and integrated optical circuits, the transfer of electromagnetic energy from source to guide, or from guide to guide, by butting the source directly up against the sink. For example butting a LED up against a fiber or fiber bundle.

Note: The input coupling coefficient by direct coupling is proportional to the square of the numerical aperture, values ranging from 0.14 to 0.50. *See also: LENS COUPLING.*

directional coupler *See TEE COUPLER.*

directive gain The ratio of 4π times the power delivered per unit solid angle (steradian) in a given direction to the power delivered to 4π steradians.

Note: 1. The directive gain is usually expressed in dB as $10 \log_{10}$ of the ratio obtained. (This yields the gain relative to an isotropic antenna.)

Note: 2. The power delivered to 4π steradians is the total power delivered by the antenna. *See also: ANTENNA; EFFECTIVE RADIATED POWER; POWER GAIN OF AN ANTENNA.*

directivity pattern A diagram relating power density (or field strength) to direction relative to the antenna, at a constant large distance from the antenna.

Note: Such diagrams usually refer to planes or the surface of a cone containing the antenna, and are usually normalized to the maximum value of the power flux density or field strength.

direct ray A ray of electromagnetic radiation that follows the path of least possible propagation time between transmitting and receiving antennas.

disc *See: OPTICAL VIDEO DISC.*

discriminator That part of a circuit which extracts the desired signal from an incoming frequency-modulated wave by changing frequency variations into amplitude variations.

disparity In PCM, the digital sum of a set of signal elements. *See also: PAIRED-DISPARITY CODE; PULSE-CODE MODULATION.*

dispersion A term used to describe the chromatic or wavelength dependence of a parameter as opposed to the temporal dependence which is referred to as distortion. The term is used, for example, to describe the process by which an electromagnetic signal is distorted because the various wavelength components of that signal have different propagation characteristics. The term is also used to describe the relationship between refractive index and wavelength.

Note: Signal distortion in an optical waveguide is caused by several dispersive mechanisms: waveguide dispersion, material dispersion, and profile dispersion. In addition, the signal suffers degradation from multimode "distortion," which is often (erroneously) referred to as multimode "dispersion." *See also: DISTORTION; INTRAMODAL DISTORTION; MATERIAL DISPERSION; MATERIAL DISPERSION PARAMETER; MULTIMODE DISTORTION; PROFILE DISPERSION; PROFILE DISPERSION PARAMETER; WAVEGUIDE DISPERSION.*

dispersion attenuation *See: OPTICAL DISPERSION ATTENUATION.*

dispersion equation An equation that indicates the dependence of the refracture index of a medium on the wavelength of the light conducted or transmitted by the medium. The adjustment of the index for wavelength permits more accurate calculation of angles or paths that are dependent upon the index. Often it is necessary to obtain a value of the rate of change of the refractive index with respect to the wavelength. The dispersion equation attributed to Hartmann is

$$N = N_0 + \frac{C}{L - L_0} ;$$

that attributed to Cauchy is

$$N = A + \frac{B}{L^2} + \frac{C}{L^4} ;$$

a more complicated one derived by Sellmeter is

$$N^2 = 1 + \sum_{i=0}^{M} \frac{A_i L^2}{L^2 - L_i^2} ;$$

An extension of the Sellmeter equation that is useful for covering more than one absorption region is

$$\sum_{i=0}^{m} \frac{A_i L^2}{L^2 - L_i^2} ;$$

finally, the Helmholtz expression which includes an additional term

$$\frac{B_i}{L^2 - L_i^2}$$

is useful within absorption regions as well. Usually, some of the terms of the summation are replaced by a constant. In practice, one of the above expressions is often used, and then a more accurate fit is found by an appropriate curve-fitting technique such as the method of least squares.

dispersion-limited operation In optical communication system operations, the situation that occurs when the dispersion of the pulse, rather than its amplitude, limits the distance between repeaters.
Note: In this regime of operation, waveguide and material dispersion are sufficient to preclude an intelligent decision based on the presence or absence of a pulse in the intelligence signal. *See also:* DETECTOR NOISE-LIMITED OPERATION.

dispersive lens *See: DIVERGING LENS.*

displacement loss *See: LATERAL DISPLACEMENT LOSS.*

dissector *See: IMAGE DISSECTOR.*

distortion A change of signal waveform shape.
Note. In a multimode fiber, the signal can suffer degradation from multimode distortion. In addition, several dispersive mechanisms can cause signal distortion in an optical waveguide: waveguide dispersion, material dispersion, and profile dispersion. *See also:* DISPERSION; PROFILE DISPERSION.

distortion-limited operation The condition prevailing when the distortion of the received signal, rather than its amplitude (or power), limits performance. The condition is reached when the system distorts the shape of the waveform beyond specified limits. For linear systems, distortion-limited operation is equivalent to bandwidth-limited operation. *See also:* ATTENUATION-LIMITED OPERATION; BANDWIDTH-LIMITED OPERATION; DISTORTION; MULTIMODE DISTORTION.

distributed Bragg-reflector laser Diode laser for which feedback of radiation via reflections is realized by periodic thickness variations of the active or cladding layers in regions outside the current-pumped region. As for DFB lasers the wavelength selectivity is high, thus allowing stable single-mode laser operation.
Note: See for examples D. Botez and G. Herskowitz, *Proc. IEEE,* vol. 68, pp. 689–732, June 1980.

distributed feedback laser Diode laser for which feedback of radiation via reflections is obtained by periodic variations in the thickness of the active or cladding layers. The reflections are in phase over a narrow wavelength range (≈ 1 nm) thus providing stable single-mode operation for diode lasers.
Note: See for examples D. Botez and G. Herskowitz, *Proc. IEEE,* vol. 68, pp. 689–732, June 1980.

distribution *See: POISSON DISTRIBUTION.*

divergence *See BEAM DIVERGENCE.*

divergent lens *See: DIVERGING LENS.*

divergent meniscus lens A lens with one convex surface and one concave surface, the latter having the greater curvature or power, the lens thus behaving generally like a concavo-concave lens, i.e., being a negative meniscus. *Synonym: DIVERGING MINISCUS LENS; NEGATIVE MENISCUS.*

diverging beam A beam of light that is not collimated, for example, one whose wavefront is spherical.
Note: A high degree of collimation, i.e. minimal divergence is required to couple energy into an optical fiber waveguide. Lasers produce beams with a high degree of collimation and uniform phase, as though the monochromatic light was emanating from a distant source.

diverging lens A lens that causes parallel light rays to spread out.
Note: One surface of a diverging lens may be concavely spherical and the other plane (planoconcave). Both may be concave (double concave) or one surface may be concave and the other convex (concave-convex, divergent-meniscus). The diverging lens is always thicker at the edge than at the center. The diverging lens is considered to have a negative focal length measured from the focal point toward the object. *Synonyms: CONCAVE LENS; DISPERSIVE LENS; DIVERGENT LENS; NEGATIVE LENS.*

diverging meniscus lens *See: DIVERGENT MENISCUS LENS.*

diversity *See DISPERSION; DIVERSITY RECEPTION.*

diversity combiner A circuit or device for combining two or more signals carrying the same information received via separate paths or channels with the objective of providing a single resultant signal which is superior in quality to any of the contributing signals. *See also: DIGITAL COMBINING; EQUAL-GAIN COMBINER; LINEAR COMBINER; MAXIMAL-RATIO COMBINER; POST-DETECTION COMBINER; PREDETECTION COMBINER; SELECTIVE COMBINER.*

diversity factor The ratio of the sum of the individual maximum demands of the various parts of a power distribution system to the maximum demand of the whole system. The diversity factor is always greater than unity. *See also: DIVERSITY RECEPTION.*

diversity reception That method of radio reception whereby, in order to minimize the effects of fading, a resultant signal is obtained by combination or selection, or both, of two or more independent sources of received-signal energy which carry the same modulation or information, but which may vary in their fading characteristics at any given instant.
Note: The amount of diversity improvement is directly dependent on the independence of the fading characteristics. *See also: DIVERSITY FACTOR; DUAL DIVERSITY; FREQUENCY DIVERSITY; ORDER OF DIVERSITY; POLARIZATION DIVERSITY; QUADRUPLE DIVERSITY; SPACE DIVERSITY; TONE DIVERSITY.*

division multiplex *See: WAVELENGTH DIVISION MULTIPLEX.*

division multiplexing *See: COLOR-DIVISION MULTIPLEXING; OPTICAL SPACE-DIVISION MULTIPLEXING.*

DoD *Acronym for: DEPARTMENT OF DEFENSE.*

DoD master clock The U.S. Naval Observatory master clock has been designated as the DoD Master Clock to which DoD time and frequency measurements are referenced (traceable).
Note: This clock is also the standard time reference for the U.S. Government in accordance with Federal Standard 1002.

dopant A material mixed, fused, amalgamated, crystallized or otherwise added to another (intrinsic) material in order to achieve desired characteristics of the resulting material. For example, the germanium tetrachloride or titanium tetrachloride used to increase the refractive index of glass for use as an optical fiber core material, or the gallium or arsenic added to silicon or germanium to produce a doped semiconductor for achieving donor or acceptor. Positive or negative impurity material for diode and transistor action.

doped-silica cladded fiber An optical fiber consisting of a doped silica core with doped silica cladding, usually produced by the Chemical Vapor Deposition (CVD) process.
Note: This fiber has a very low loss and moderate dispersion. It is a step-indexed fiber.

doped-silica graded fiber An optical fiber consisting of a silica fiber in which the doping varies so as to produce a decreasing refractive index from the center toward the outside, thus eliminating the necessity of cladding.
Note: The refractive index profile is graded and tailored to reduce multimode dis-

persion. Since the non-axial rays of light, though traveling further, travel faster in the outer medium where the refractive index is lower. Thus the axial rays arrive at the end of the fiber the same time as the non-axial or paraxial rays.

doping concentration The relative amount of dopant atoms in a semiconductor or glass material. For semiconductors the doping concentration is expressed in number of atoms per cubic centimeter. For glasses the doping or impurity concentration is expressed in percentages or ppm.

Doppler effect The phenomenon evidenced by the change in the observed frequency of a sound or radio wave caused by a time rate of change in the magnitude of the radial component of relative velocity between the source and the point of observation.

double crucible method A method of fabricating an optical waveguide by melting core and clad glasses in two suitably joined concentric crucibles and then drawing a fiber from the combined melted glass. *See also: CHEMICAL VAPOR DEPOSITION TECHNIQUE.*

double-ended control *Synonym: DOUBLE-ENDED SYNCHRONIZATION.*

double-ended synchronization A synchronization control system between two exchanges in which the phase error signals used to control the clock at one exchange are derived from comparison of the phase of the incoming digital signal and the phase of the internal clock at both exchanges. *Synonym: DOUBLE-ENDED CONTROL. See also: SYNCHRONIZATION.*

double frequency-shift keying (DFSK) A multiplex system in which two telegraph signals are combined and transmitted simultaneously by a method of frequency shifting among four radio frequencies.

double heterojunction In a laser diode, two heterojunctions in close proximity, resulting in full carrier and radiation confinement, and thus improved control of recombinations and waveguiding.

double heterojunction diode A laser diode that has two different heterojunctions, the difference being primarily in the stepped changes in refractive index and bandgap energy of the material in the vicinity of the p-n junction. Symbol: DH diode.

Note: The double heterojunction laser diode is widely used for pulse-code (CW) operation.

double-heterostructure *See: DOUBLE-HETEROJUNCTION.*

double-image Pertaining to the doubling of an image caused by optical imperfections in the optical system.

double-sideband suppressed carrier transmission That method of transmission in which the frequencies produced by the process of amplitude modulation are symmetrically spaced both above and below the carrier. The carrier level is suppressed to a predetermined value below the level of the transmitted sidebands. *See also: CARRIER; SIDEBAND TRANSMISSION.*

double-sideband transmission That method of sideband transmission in which both sidebands are transmitted.

doublet In optics, a compound lens consisting of two elements.

Note: If there is an air space between the elements it is called an air-spaced doublet. If the inner surfaces are cemented together, it is called a cemented doublet.

doubly-refracting crystal A transparent crystalline substance that is anisotropic with respect to the velocity of light traveling within it in two different directions, i.e. with respect to its refractive index in two different directions.

down-converter A type of converter which is characterized by the frequency of the output signal being lower than the frequency of the input signal. It is the converse of *UP-CONVERTER.*

downlink That portion of a communication link used for transmission of signals from a satellite or airborne platform to a surface terminal. It is the converse of *UPLINK. See also: SATELLITE.*

DPSK *Acronym for DIFFERENTIAL PHASE-SHIFT KEYING.*

D region *See IONOSPHERE.*

drift current Current set by carriers in an electric field. Of importance in photodetectors and to a lesser extent in carrier leakage from the active layer of diode lasers and LEDs.

drive circuit In optical fiber transmission systems, the electrical circuit that drives the light-emitting source, modulating it in accordance with an intelligence-bearing signal.

DSE *Acronym for DATA SWITCHING EXCHANGE*

D-star *See D*.*

DSTE *Acronym for DATA SUBSCRIBER TERMINAL EQUIPMENT.*

DTE *Acronym for DATA TERMINAL EQUIPMENT.*

DTE clear signal A call control signal sent by the DTE to initiate clearing. *See also: DATA TERMINAL EQUIPMENT.*

DTE waiting signal A call control signal at the DCE/DTE interface that indicates that the DTE is waiting for a call control signal from the DCE. *See also: DATA TERMINAL EQUIPMENT.*

DTMF *Acronym for DUAL-TONE MULTI-FREQUENCY SIGNALING.*

dual access 1. The connection of a user or subscriber to two switching centers by separate access lines using a single message routing indicator or telephone number. 2. In satellite communications, the transmission of two carriers simultaneously through a single communications satellite repeater. *See also: MULTIPLE ACCESS.*

dual diversity The simultaneous combining of, or selection from, two independently fading signals and their detection through the use of space, frequency, angle, time, or polarization characteristics. *See also: DIVERSITY RECEPTION; FREQUENCY DIVERSITY; ORDER OF DIVERSITY; QUADRUPLE DIVERSITY.*

dual-tone multifrequency signaling (DTMF) A telephone signaling method employing set combinations of two specific voice-band frequencies, one of which is selected from a group of four low frequencies, and the other from a group of either three or four relatively high frequencies.

Note: It is used by subscribers and PBX attendants, if their switchboard positions are so equipped, to indicate telephone address digits, precedence ranks, and end-of-signaling. Civil telephones using DTMF normally have 12 key combinations, the ten digits plus # and *, the latter being reserved for special purposes. AUTOVON telephones have 16 combinations, the extra 6 being used for precedence, etc. DTMF signals, unlike dial pulses, can pass through the entire connection to the called party, and therefore lend themselves to various schemes for remote control, etc., after the connection is set up. *See also: KEY PULSING.*

duct *See ATMOSPHERIC DUCT.*

ducting The propagation of radio waves within an atmospheric duct.

duplex circuit A circuit that affords simultaneous operation in opposite directions. *See also: DUPLEX OPERATION.*

duplexer A device that permits the simultaneous use of a transmitter and a receiver in connection with a common element such as an antenna system. *See also: DUPLEX OPERATION.*

duplex operation A type of operation in which simultaneous two-way conversations, messages, or information may be passed between any two given points. *Synonyms: FULL-DUPLEX OPERATION; TWO-WAY SIMULTANEOUS OPERATION. See also: DIPLEX OPERATION; DUPLEX CIRCUIT; DUPLEXER; HALF-DUPLEX OPERATION; SEMI-DUPLEX OPERATION; SIMPLEX OPERATION.*

duty cycle In pulsed device operation the product between pulse width and the pulse repetition rate.

dynamic range 1. In a transmission system, the difference in decibels between the noise level of the system and its overload level. 2. The difference, in decibels, between the overload level and the minimum acceptable signal level in a system or transducer.

dynamic scanning In optical fiber transmission systems, a technique in which a fiber bundle is vibrated about a fixed point with reference to the impressed image in order to suppress the fiber pattern, i.e. render the fiber pattern less visible at the output end.

dynamic variation (transient) Short time variations outside of steady state conditions in the characteristics of power delivered to the communications equipment.

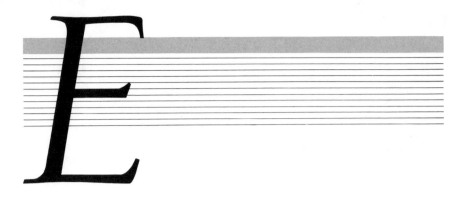

earth coverage In satellite communications, the condition obtained when a beam is sufficiently wide to cover the surface of the earth exposed to the satellite. *See also: FOOTPRINT; SATELLITE.*

echo A wave which has been reflected or otherwise returned with sufficient magnitude and delay to be perceived.
Note: Echoes are frequently measured in dB relative to the directly transmitted wave.

echo attenuation In a four-wire (or two-wire) circuit in which the two directions of transmission can be separated from each other, the attenuation of the echo signals (which return to the input of the circuit under consideration) is determined by the ratio of the transmitted power to the echo power received; expressed in dB.

echo check A method of checking the accuracy of transmission of data, by which the received data are returned to the sending end for comparison with the original data. *See also: INFORMATION FEEDBACK.*

echoplex An echo check applied to network terminals operating in the two-way simultaneous mode. *See also: NETWORK.*

echo suppressor A voice-operated device for connection to a two-way telephone circuit to attenuate echo signals in one direction caused by telephone signals in the other direction.

EDC switch *See: ELECTRO-OPTIC DIRECTIONAL COUPLER SWITCH.*

edge-emitting diode *See: RESTRICTED EDGE-EMITTING DIODE.*

edge-emitting LED A light emitting diode with a special output that emanates from between the heterogeneous layers, i.e. from an edge, having a higher radiance and greater coupling efficiency to an optical fiber or integrated optical circuit than the surface-emitting LED, but not as great as the injection laser.
Note: Surface-emitting and edge-emitting LEDs provide several milliwatts of power in the 0.8-1.5 micron spectral range at drive currents of 100–200 milliamperes; diode lasers at these currents provide tens of milliwatts. *See also: SURFACE-EMITTING LED.*

edge-response The ability of an optical fiber bundle to form, maintain, and resolve an image of a sharply-outlined image, i.e. a knife-edge.

edge test *See: FOUCAULT KNIFE-EDGE TEST.*

effect *See ACOUSTO-OPTIC EFFECT, ELECTRO-OPTIC EFFECT; MAGNETO-OPTIC EFFECT; PHOTOCONDUCTIVE EFFECT; PHOTOELECTRIC EFFECT; PHOTOELECTROMAGNETIC EFFECT; PHOTOEMISSIVE EFFECT; PHOTOVOLTAIC EFFECT; STARK EFFECT; ZEEMAN EFFECT.*

effective antenna length The ratio of the open-circuit voltage of an antenna to the electric field intensity. *See also: ANTENNA.*

effective data transfer rate The average number of bits, characters, or blocks per unit time transferred from a data source and accepted as valid by a data sink. It is expressed in bits, characters, or blocks per second, minute, or hour.

effective earth radius The radius of a hypothetical earth for which the distance to the radio horizon, assuming rectilinear propagation, is the same as that for the actual earth with an assumed uniform vertical gradient of refractive index.
Note: For the standard atmosphere, the effective radius is 4/3 that of the actual earth. *See also: FRESNEL ZONE; K-FACTOR; PATH CLEARANCE; PROPAGATION PATH OBSTRUCTION.*

effective height 1. The height of the center of radiation of an antenna above the effective ground level. 2. In low frequency applications, involving loaded or nonloaded vertical antennas, the moment of the current distribution in the vertical section divided by the input current.

Note: For an antenna with symmetrical current distribution, the center of radiation is the center of distribution. For an antenna with asymmetrical current distribution, the center of radiation is the center of current moments when viewed from directions near the direction of maximum radiation. *See also: ANTENNA.*

effective input noise temperature The source noise temperature in a two-port network or amplifier that will result in the same output noise power, when connected to a noise-free network or amplifier, as that of the actual network or amplifier connected to a noise-free source.

Note: If F is the noise figure and 290 K the standard noise temperature, then the effective noise temperature, T_n, is:

$$T_n = 290(F-1).$$

See also: NOISE.

effective mode volume The square of the product of the diameter of the near-field pattern and the sine of the radiation angle of the far-field pattern. The diameter of the near-field radiation pattern is defined here as the full width at half maximum and the radiation angle at half maximum intensity.

Note: Effective mode volume is proportional to the breadth of the relative distribution of power amongst modes in a multimode fiber. It is not truly a spatial volume but rather an "optical volume" equal to the product of area and solid angle. *See also: MODE VOLUME; RADIATION PATTERN.*

effective radiated power (ERP) The power supplied to the antenna multiplied by the power gain of the antenna in a given direction. *See also: ANTENNA; DIRECTIVE GAIN.*

effective refractive index A quantity characterizing the apparent wavelength seen by an optical mode propagating through the dielectric waveguide of a fiber or of a semiconductor optoelectronic device. The ratio between the mode propagation constant and the free-space wavenumber. Of importance in the design of single-mode fibers, mode-stabilized diode lasers and DFB and DBR lasers.

Note: See for example H. Kogelnik, *IEEE Trans. Microwave Theory Tech.,* vol. MTT-23, p. 1, Jan. 1975.

effective speed of transmission The rate at which information is processed by a transmission facility expressed as the average rate over some significant time interval. This quantity is usually expressed as the average number of characters or bits per unit time. *Synonym: AVERAGE RATE OF TRANSMISSION. See also: EFFICIENCY FACTOR; THROUGHPUT.*

efficiency factor In telegraph communications, the ratio of the time to transmit a text automatically and at a specified modulation rate, to the time actually required to receive the same text with a specified error rate.

Note: 1. All of the communication facilities are assumed to be in the normal condition of adjustment and operation.

Note: 2. Telegraph communications may have different temporal efficiency factors for the two directions of transmission.

Note: 3. The practical conditions of measurement should be specified; in particular, the duration. *See also: EFFECTIVE SPEED OF TRANSMISSION.*

EFL *Acronym for EQUIVALENT FOCAL LENGTH.*

E-layer *See IONOSPHERE.*

electrical bandwidth For light-emitting diodes, the frequency at which the optical output drops by 1.5 dB with respect to its low-frequency value.

electrical contact For semiconductor devices, the metallization needed to pass current from a soldered wire lead or heat-sink to the device.

electroluminescence Nonthermal conversion of electrical energy into light. One example is the photon emission resulting from electron-hole recombination in a pn junction such as in a light emitting diode. *See also: INJECTION LASER DIODE.*

electromagnetic interference (EMI) The phenomenon resulting when electromagnetic energy causes an unacceptable or undesirable response, malfunction, degradation, or interruption of the intended operation of an electronic equipment, subsystem, or system. *Synonym: RADIO FREQUENCY INTERFERENCE (RFI). See also: ELECTROMAGNETIC INTERFERENCE CONTROL; INTERFERENCE; INTERFERENCE EMISSION.*

electromagnetic radiation Radiation made up of oscillating electric and magnetic fields and propagated with the speed of light. Includes gamma radiation, X-rays, ultraviolet, visible and infrared radiation, and radar and radio waves. *See also: RADIATION PATTERN; RADIATION SCATTERING.*

electromagnetic spectrum 1. The entire range of wavelengths, extending from the shortest to the longest or conversely, that can be generated physically. *Note:* This range of electromagnetic wavelengths extends almost from zero to infinity and includes the visible portion of the spectrum known as light.
2. The frequencies (or wave lengths) present in a given electromagnetic radiation. A particular spectrum could include a single frequency or a wide range of frequencies.

electromagnetic theory The theory of propagation of energy by combined electric and magnetic fields. *Note:* Much of the theory is embodied in Maxwell's equations.

electron optics The area of science devoted to the directing and guiding of electron beams using electric fields in the same manner as lenses are used on light beams; for example, in image-converter tubes and electron microscopes, pertaining to devices whose operation relies on modification of a material's refractive index by electric fields. *Note:* In a Kerr cell, the index change is proportional to the square of the electric field, and the material is usually a liquid. In a Pockel cell, the material is a crystal whose index change is linear with the electric field.

electronic charge The quantity of charge represented or possessed by one electron, equal to 1.6×10^{-19} coulomb.

electronic device *See: OPTOELECTRONIC DEVICE.*

electro-optic coefficient A measure of the extent to which the index of refraction changes with applied high electric field, such as several parts per ten thousand for applied fields of the order of 20 volts per centimeter. *Note:* Since the phase shift of a light wave is a function of the index of refraction of the medium in which it is propagating, the change in index can be used to phase-modulate the light wave by shifting its phase at a particular point along the guide, by changing the propagation time to the point. *See also: ELECTRO-OPTIC PHASE MODULATION.*

electro-optic directional coupler (EDC) switch Guided-wave optics device in which by application of voltages, controlled transfer of energy is obtained between two optical waveguides. Thus light could be totally switched from one waveguide to another. The process is possible since the applied electric fields locally change the index of refraction of the material (e.g., $LiNbO_3$) via the electro-optic effect. *Note:* See for examples D. Botez and G. Herskowitz, *Proc. IEEE*, vol. 68, pp. 689–732, June 1980.

electro-optic effect A change in the refractive index of a material under the influence of an electric field. *Note 1.* Pockels and Kerr effects are electro-optic effects that are respectively linear and quadratic in the electric field strength. *Note 2.* Electro-optic is often erroneously used as a synonym for optoelectronic. *See also: OPTOELECTRONIC.*

electro-optic phase modulation Modulation of the phase of a lightwave, such as by changing the index of refraction and thus the velocity of propagation and hence the phase at a point in the medium in which the wave is propagating, in accordance with an applied field serving as the modulating signal. *See also: ELECTRO-OPTIC COEFFICIENT.*

element *See: RECEIVING ELEMENT; TRANSMITTING ELEMENT.*

emergence Pertaining to the trigonometric relation between an emergent ray and the surface of a medium. *See also: GRAZING EMERGENCE; NORMAL EMERGENCE.*

emergent ray A ray of light leaving, i.e., emerging from a medium as contrasted to an entering or incident ray.

EMI *Acronym for ELECTROMAGNETIC INTERFERENCE.*

emission-beam-angle-between-half-power-points The angle centered on the optical axis of a light-emitter within which the radiant power density is equal to or greater than half the maximum power density (on the optical axis).

emission law *See:* COSINE EMISSION LAW.

emission of radiation *See:* MICRO-WAVE AMPLIFICATION BY STIMU-LATED EMISSION OF RADIATION.

emissivity The ratio of power radiated by a substance to the power radiated by a blackbody at the same temperature. Emissivity is a function of wavelength and temperature. *See also:* BLACKBODY.

emittance *See:* LUMINOUS EMIT-TANCE; RADIANT EMITTANCE; SPEC-TRAL EMITTANCE.

emitter *See:* OPTICAL EMITTER.

emitting diode *See:* LIGHT EMITTING DIODE; RESTRICTED EDGE-EMITTING DIODE.

emitting diode coupler *See:* LIGHT-EMITTING DIODE COUPLER.

emitting LED *See:* EDGE-EMITTING LED; SURFACE-EMITTING LED.

emphasis The intentional alteration of the frequency-amplitude characteristics of a signal to reduce adverse effects of noise in a communication system.
Note: An example of this is the pre-emphasis used in the transmission of a frequency modulated wave. The higher frequency signals are emphasized in order to produce a more equal modulation index for the transmitted frequency spectrum and therefore a better signal-to-noise ratio for the entire frequency range. *See also:* DE-EMPHASIS; PRE-EM-PHASIS.

EMS *Acronym for* EQUILIBRIUM MODE SIMULATOR.

encode 1. To apply a code, frequently one consisting of binary numbers, to represent individual characters or groups of characters in a message. 2. To substitute letters, numbers, or characters for other numbers, letters, or characters, usually intentionally to hide the meaning of the message except to certain persons who know the encoding scheme. 3. To convert plain-text into unintelligible form by means of a code-system. *See also:* CODE; DECODE; ENCRYPT.

encoder *See:* ANALOG-TO-DIGITAL EN-CODER.

encoding *See:* ANALOG ENCODING.

encoding law The law defining the relative values of the quantum steps used in quantizing and encoding signals. *See also:* SEGMENTED ENCODING LAW.

end-finish *See:* OPTICAL END-FINISH.

end-fire coupling Optical fiber and in-tegrated optical-circuit (IOC) coupling between two waveguides in which the two waveguides to be coupled are butted up against each other. A more straight-forward, simpler, and more efficient coupling method than evanescent field coupling.
Note: Mode pattern matching is required and accomplished by maintaining a uni-ty cross-sectional area aspect ratio, axial alignment, and minimal lateral axial dis-placement. *See also:* EVANESCENT FIELD COUPLING.

endoscope An optical instrument used to obtain images of, or view, internal parts of a system, such as internal cavi-ties and organs of the body.
Note: Endoscopes often use optical fiber bundles as transmission cables.

energy *See:* RADIANT ENERGY.

energy density *See:* OPTICAL ENERGY DENSITY.

energy level The discrete amount (quanta) of kinetic and potential energy possessed by an orbiting electron.
Note: Energy (quanta) is absorbed or ra-diated depending on whether an electron moves from a lower to a higher or a higher to a lower energy level.

entrance pupil 1. In an optical system, the image of the limiting aperture stop formed in the object space by all optical elements of the system preceding the limiting aperture stop. 2. The aperture of the objective when there are no other limiting stops following it in an optical system.

envelope 1. The boundary of the family of curves obtained by varying a parame-ter of a wave. 2. A group of binary digits formed by a byte augmented by a num-ber of additional bits which are required for the operation of the data network. *See also:* INFORMATION BEARER CHAN-NEL; ISOCHRONOUS BURST TRANS-MISSION; PACKET.

envelope delay distortion In a given passband of a device or a transmission facility, the maximum difference of the group delay time between any two spec-ified frequencies. *See also:* GROUP DE-LAY TIME.

epitaxy Technique of single-crystal growth by chemical reaction used to form, on the surface of a crystalline sup-port substrate, thin layers of material with lattice structures identical to that of the substrate. Origin: "epi" meaning on in Greek and "taxis" meaning arrange-ment in Greek. *See:* LIQUID PHASE EPI-

TAXY; VAPOR PHASE EPITAXY; MO-LECULAR BEAM EPITAXY.

equal gain combiner A diversity combiner in which the signals on each channel are added together; the channel gains are all equal and can be made to vary equally so that the resultant signal is approximately constant. *See also: DIVERSITY COMBINER; GAIN.*

equalization The process of reducing frequency distortion or phase distortion, or both, of a circuit by the introduction of networks to compensate for the difference in attenuation, time delay, or both, at the various frequencies in the transmission band. *See also: SPECIAL GRADE ACCESS LINE.*

equalizer *See: AMPLITUDE EQUALIZER; DELAY EQUALIZER; SLOPE EQUALIZER.*

equations *See: DISPERSION EQUATIONS.*

equilibrium coupling length *Synonym: EQUILIBRIUM LENGTH.*

equilibrium length For a specific excitation condition, the length of multimode optical waveguide necessary to attain equilibrium mode distribution. *Note:* The term is sometimes used to refer to the longest such length, as would result from a worst-case, but undefined excitation. *Synonyms: EQUILIBRIUM COUPLING LENGTH; EQUILIBRIUM MODE DISTRIBUTION LENGTH. See also: EQUILIBRIUM MODE DISTRIBUTION; MODE COUPLING.*

equilibrium mode distribution The condition in a multimode optical waveguide in which the relative power distribution among the propagating modes is independent of length. *Synonym: STEADY-STATE CONDITION. See also: EQUILIBRIUM LENGTH; MODE; MODE COUPLING.*

equilibrium mode distribution length *Synonym: EQUILIBRIUM LENGTH.*

equilibrium mode simulator (EMS) A device or optical system used to create an approximation of the equilibrium mode distribution. *See also: EQUILIBRIUM MODE DISTRIBUTION; MODE FILTER.*

equivalent focal length (EFL) 1. The distance from a principal point to its corresponding principal focal point. 2. The focal length of the equivalent thin lens. *Note:* The size of the image of an object is directly proportional to the equivalent focal length of the lens forming it.

equivalent noise resistance A quantitative representation in resistance units of the spectral density of a noise voltage generator at a specified frequency. *Note: 1.* The relation between the equivalent noise resistance R_n and the spectral density W_n of the noise-voltage generator is $R_n = \pi W_n / kT_0$ where k is Boltzmann's constant and T_0 is the standard noise temperature (290 K) and $kT_0 = 4.00 \times 10^{-21}$ watt-seconds. *Note: 2.* The equivalent noise resistance in terms of the mean-square noise-generator voltage e^2 within a frequency increment Δf is $R_n = e^2 / 4kT_0 \Delta f$. *See also: NOISE.*

equivalent PCM noise Through comparative tests, the amount of thermal noise power on an FDM or wire channel necessary to approximate the same judgment of speech quality created by quantizing noise in a PCM channel. Generally, 33.5 dBrnC ±2.5 dB is considered the approximate equivalent PCM noise of a 7-bit PCM system. *See also: PULSE-CODE MODULATION.*

equivalent power *See: NOISE EQUIVALENT POWER.*

erect image In an optical system, an image, either real or virtual, that has the same spatial orientation as the object. *Note:* The image obtained at the retina with the assistance of an optical system is said to be erect when the orientation of the image is the same as with the unaided eye.

erect position In frequency division multiplexing, a position of a translated channel in which an increasing signal frequency in the untranslated channel causes an increasing signal frequency in the translated channel. *Synonym: UPRIGHT POSITION.*

erlang An international (dimensionless) unit of the average traffic intensity (occupancy) of a facility during a period of time, normally a busy hour. The number of erlangs is the ratio of the time during which a facility is occupied (continuously or cumulatively) to the time this facility is available for occupancy. *Note: 1.* The facility, such as a switch register, trunk, circuit, etc, is usually shared by many users. *Note: 2.* Communications traffic measured in erlangs and offered to a group of shared facilities, such as a trunk group, is equal to the sum of the traffic intensity (in erlangs) of all individual

sources (e.g., telephones) that share and are served exclusively by this group of facilities.
Note: 3. A single facility (register, trunk, circuit, etc.) cannot carry more than one erlang of traffic. *Synonym: TRAFFIC UNIT. See also: BUSY HOUR; CALL-SECOND; GROUP BUSY HOUR; NETWORK BUSY HOUR; SWITCH BUSY HOUR; TRAFFIC INTENSITY; TRAFFIC LOAD.*

ERP *Acronym for EFFECTIVE RADIATED POWER.*

error 1. The difference between a computed, estimated, or measured value and the true, specified, or theoretically correct value. 2. A malfunction that is not reproducible. *See also: BIT ERROR RATE; DIGITAL ERROR; ERROR BURST; FAULT; FORWARD ERROR CORRECTION; RESIDUAL ERROR RATE.*

error burst A group of bits in which two successive erroneous bits are always separated by less than a given number (M) of correct bits.
Note: The last erroneous bit in a burst and the first erroneous bit in the following burst are accordingly separated by M correct bits or more. The number M should be specified when describing an error burst. *See also: BURST; ERROR.*

error control The improvement of digital communications through use of certain techniques such as error detection, forward-acting error correction, and use of block codes.

error-correcting code A code in which each telegraph or data signal conforms to specific rules of construction so that departures from this construction in the received signals can be automatically detected. This permits the automatic correction, at the receiving terminal, of some or all of the errors. Such codes require more signal elements than are necessary to convey the basic information.

error-correcting system A system employing an error-correcting code and so arranged that some or all signals detected as being in error are automatically corrected at the receiving terminal before delivery to the data sink. *See also: COMMUNICATIONS SYSTEM.*

error - detecting - and - feedback system *Synonym: AUTOMATIC REQUEST-REPEAT (ARQ).*

error-detecting code A code in which each telegraph or data signal conforms to specific rules of construction, so that departures from this construction in the received signals can be automatically detected. Such codes require more signal elements than are necessary to convey the basic information.

error-detecting system A system employing an error-detecting code and so arranged that any signal detected as being in error is either deleted from the data delivered to the data sink, in some cases with an indication that such deletion has taken place, or delivered to the data sink together with an indication that the signal is in error.

error rate The ratio of the number of bits, elements, characters, or blocks incorrectly received to the total number of bits, elements, characters or blocks sent in a specified time interval. *See also: BIT ERROR RATE (BER).*

evanescent field A time varying electromagnetic field whose amplitude decreases monotonically, but without an accompanying phase shift, in a particular direction is said to be evanescent in that direction.

evanescent-field coupling Optical fiber, or integrated optical-circuit (IOC) coupling between two waveguides in which the two waveguides to be coupled are held parallel to each other in the coupling region. The evanescent waves on the outside of the waveguide entering the coupled waveguide bringing some of the light energy with it into the coupled waveguide.
Note: Close-to-core proximity or fusion is required. *See also: TANGENTIAL COUPLING.*

exalted-carrier reception A method of receiving either amplitude or phase modulated signals in which the carrier is separated from the sidebands, filtered and amplified, and then combined with the sidebands again at a higher level prior to demodulation.

excess insertion loss In an optical waveguide coupler, the optical loss associated with that portion of the light which does not emerge from the nominally operational ports of the device. *See also: OPTICAL WAVEGUIDE COUPLER.*

excess noise factor In avalanche photodiodes, a measure of the increase in shot noise compared to an ideal noiseless multiplier. Excess noise is created due to statistical variations in avalanche gain, M. Of importance in detectors made of semiconductors with similar ionization coefficient values for electrons and holes; such as, germanium and III-V compound photodiodes. Symbol: F. F increases with multiplication, causing the noise power to increase faster than the signal power, and then a maximum signal-to-noise ratio is obtained at a fixed gain value.

Note: See for examples S. M. Sze, *Physics of Semiconductor Devices,* 2nd edition, John Wiley & Sons, New York, 1981.

exchange process *See: DOUBLE-CRUCIBLE PROCESS.*

excited atmosphere laser *See: LONGITUDINALLY-EXCITED ATMOSPHERE LASER; TRANSVERSE-EXCITED ATMOSPHERE LASER.*

excited state Any orbital kinetic and potential energy state that an electron can have above the ground state. *See also: GROUND STATE.*

Note: An electron emits a quantum of energy when it moves from an excited state to the ground state.

exitance *See: RADIANT EXITANCE.*

exit angle When a light ray emerges from a surface, the angle between the ray and a normal to the surface at the point of emergence.

Note: For an optical fiber, the angle between the output ray, and the axis of the fiber or fiber bundle.

exit pupil In an optical system, the image of the limiting aperture stop formed by all lenses following this stop.

Note: In photographic objectives, this image is virtual and is usually not far from the iris diaphragm; in telescopes, the image is real and can be seen as a small, bright, circular disc by looking at the eyepiece of the instrument directed toward an illuminated area or light source; in telescopes, its diameter is equal to the diameter of the entrance pupil divided by magnification of the instrument; in Galilean telescopes, the exit pupil is a virtual image between the objective and eyepiece and acts as an out-of-focus field stop.

expandor A device with a nonlinear gain characteristic that acts to increase the gain more on larger input signals than it does on smaller input signals.

Note: Usually used to restore the original dynamic range of a signal that was compressed for transmission. *See also: CLIPPER; COMPANDOR; COMPRESSOR; PEAK LIMITING.*

external differential quantum efficiency The rate of increase of the optical photon flux (i.e., power over photon energy) with increasing current in the lasing mode of a diode laser. Also referred to as slope efficiency. Symbol: η_{ext}.

external optical modulation Modulation of a light wave in a medium by application of fields, forces, waves, or other energy forms upon a medium conducting a light beam in such a manner that a characteristic of either the medium, or the beam, or both, are modulated in some fashion.

Note: External optical modulation can make use of such effects as the electro-optic acousto-optic, magneto-optic, or absorptive effect.

external photoeffect The emission of photon-excited electrons from the surface of a material after overcoming the energy barrier at the surface of a photoemissive material.

external photoeffect detector A photodetector in which the energy of each photon incident on the detector surface is sufficient to liberate one or more electrons, i.e., Planck's constant times the frequency, which is the energy of the photon, is sufficient to overcome the work function of the material, and the liberated electrons move in under the influence of an applied electric field.

Note: Photo-emissive devices make use of the external photoeffect. *See also: INTERNAL PHOTOEFFECT DETECTOR.*

external quantum efficiency The ratio between the output optical photon flux (i.e., power over photon energy) and the input current of a light emitting diode.

extinction coefficient The sum of the absorption coefficient and the scattering coefficient. *See also: BOUGER'S LAW.*

extramural absorption The absorption of light, transmitted radially through the cladding of an optical fiber, by means of a dark or opaque coating placed over the cladding.

Note: Extramural absorption may be accomplished by any light-absorbing material, such as secondary coatings, interstitial fibers, or other jackets.

extramural cladding A layer of dark or opaque absorbing coating placed over the cladding of an optical fiber to increase internal reflection, protect the smooth reflecting wall of the cladding, and absorb scattered or escaped stray light that might penetrate the cladding.

extraordinary ray   A light ray that has a nonisotropic velocity in a doubly refracting crystal.

Note: An extraordinary ray does not necessarily obey Snell's Law upon refraction at the crystal surface.

extrinsic internal photoeffect   An internal photoeffect involving the dopants or other impurities in a basic (intrinsic) material. *See also: INTRINSIC INTERNAL PHOTOEFFECT.*

extrinsic joint loss   That portion of joint loss that is not intrinsic to the fibers (i.e., loss caused by imperfect jointing). *See also: ANGULAR MIS-ALIGNMENT LOSS; GAP LOSS; INTRINSIC JOINT LOSS; LATERAL OFF-SET LOSS.*

eye pattern   An oscilloscope pattern used for examination of digital signal distortion.

Note: An open eye pattern corresponds to minimal signal distortion. Distortion of the signal waveform due to intersymbol interference and noise appears as closure of the eye pattern.

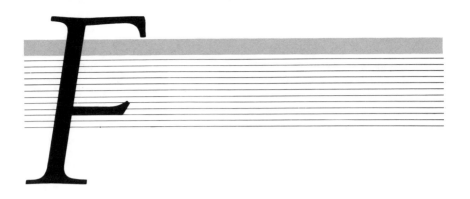

F1A-line weighting A noise weighting used in a noise measuring set to measure noise on a line that would be terminated by a 302-type or similar instrument. The meter scale readings are in dBa(F1A).
Note: F1A-line weighting is obsolete for DoD applications.

Fabry-Perot cavity Optical resonator structure for which feedback is provided by two parallel planes. For diode lasers parallel planes are obtained by cleaving, etching, or polishing.

Fabry-Perot interferometer A high-resolution multiple-beam interferometer consisting of two optically flat and parallel glass or quartz plates held a short fixed distance apart, the adjacent surfaces of the plates or interferometer flats being made almost totally reflecting by a thin silver film or multilayer dielectric coating.

Fabry-Perot laser Laser oscillator composed of a Fabry-Perot cavity in which the space between the two mirrors contains an amplifying medium with an inverted atomic population. Conventional diode lasers are said to be of the Fabry-Perot type to distinguish them from DFB, DBR, or ring types of semiconductor lasers.

faceplate *See: FIBER FACEPLATE.*

facet degradation In diode lasers or LEDs, the ensemble of degradation mechanisms that occurs at the mirror facets. At low output power levels facet degradation is slow and consists of oxidation and/or chemical erosion induced optically or by water vapor. A method of prevention is coating the facets with half-wave (passivation) films. At high power levels facet damage is thermally induced due to intense laser-light absorption at the semiconductor-air interface. For AlGaAs devices catastrophic optical damage is initiated at the mirror facets. This can be prevented by using the non-absorbing-mirror (NAM) approach.

factor *See: PHOTOCONDUCTIVE GAIN FACTOR; PULSE DUTY-FACTOR.*

fade margin 1. An allowance provided in system gain (or sensitivity) to accommodate expected fading to insure the required grade of service will be maintained for the specified percentage of time. 2. The amount by which a received signal level may be reduced without causing the system (or channel) output to fall below a specified threshold. *Synonym: FADING MARGIN. See also: FADING; RF POWER MARGIN.*

fading The variation, with time, of the intensity or relative phase, or both, of any or all frequency components of a received signal due to changes in the characteristics of the propagation path with time. *See also: FADE MARGIN; FADING DISTRIBUTION; FLAT FADING; PHASE INTERFERENCE FADING; RAYLEIGH FADING; SELECTIVE FADING.*

fading distribution The probability that signal level fading will exceed a certain value relative to a certain reference level.
Note: In the case of phase interference fading, the time distribution of the instantaneous field intensity approximates a Rayleigh distribution a large part of the time when at least several signal components of equal amplitude are involved. *See also: RAYLEIGH FADING.*

Faraday effect *See: MAGNETO-OPTIC EFFECT.*

far-end crosstalk Crosstalk which is propagated in a disturbed channel in the same direction as the propagation of a signal in the disturbing channel. The terminals of the disturbed channel at which the far-end crosstalk is present and the energized terminals of the disturbing channel are usually remote from each other. *See also: CROSS-TALK.*

far field *Synonym: FAR FIELD REGION.*

far-field diffraction pattern The diffraction pattern of a source (such as an LED, ILD, or the output end of an optical waveguide) observed at an infinite distance from the source. Theoretically, a far-field pattern exists at distances that are large compared with s^2/λ, where s is a characteristic dimension of the source and λ is the wavelength. Example: If the source is a uniformly illuminated circle, then s is the radius of the circle.

Note: The far-field diffraction pattern of a source may be observed at infinity or (except for scale) in the focal plane of a well-corrected lens. The far-field pattern of a diffracting screen illuminated by a point source may be observed in the image plane of the source. *Synonym: FRAUNHOFER DIFFRACTION PATTERN. See also: DIFFRACTION; DIFFRACTION LIMITED.*

far-field pattern *Synonym: Far-field radiation pattern.*

far-field region The region, far from a source, where the diffraction pattern is substantially the same as that at infinity. *See also: FAR-FIELD DIFFRACTION PATTERN.*

far infrared Pertaining to electromagnetic wavelengths from 30 to 1000 micrometers.

far zone *Synonym: FAR FIELD REGION.*

FDHM *Acronym for FULL DURATION AT HALF MAXIMUM. See also: FULL WIDTH (DURATION) HALF MAXIMUM.*

FDM *Acronym for FREQUENCY DIVISION MULTIPLEX.*

FEC *Acronym for FORWARD ERROR CORRECTION.*

feedback 1. The return of a portion of the output of a device to the input; positive feedback adds to the input, negative feedback subtracts from the input. 2. Information returned as a response to an originating source. *See also: INFORMATION FEEDBACK.*

feeder echo noise Distortion of a signal as a result of reflected waves in a transmission line that is many wavelengths long and mismatched at both generator and load ends.

fep-clad silica fiber *See: LOW-LOSS FEP-CLAD SILICA FIBER.*

Fermat principle A ray of light from one point to another, including reflections and refractions that may occur, follows the path that requires the least time.

Note: The optical path length is an extreme path in the terminology of the calculus of variations. *Synonym: LEAST-TIME PRINCIPLE.*

ferrule A mechanical fixture, generally a rigid tube, used to confine the stripped end of a fiber bundle or a fiber. *See also: FIBER BUNDLE.*

Note 1. Typically, individual fibers of a bundle are cemented together within a ferrule of a diameter designed to yield a maximum packing fraction. *See also: PACKING FRACTION.*

Note 2. Nonrigid materials such as shrink tubing may also be used for ferrules for special applications. *See also: REFERENCE SURFACE.*

FET *Acronym for FIELD-EFFECT TRANSISTOR.*

FET photodetector A photodetector employing photogeneration of carriers in the channel region of an FET structure to provide photodetection with current gain. *See also: PHOTOCURRENT; PHOTODIODE.*

FFL *Acronym for FRONT FOCAL LENGTH.*

fiber *See: OPTICAL FIBER.*

fiber absorption In an optical fiber, the lightwave power attenuation due to absorption in the fiber core material, a loss usually evaluated by measuring the power emerging at the end of successively-shortened known lengths of the fiber.

fiber axis The line connecting the centers of the circles that circumscribe the core, as defined under Tolerance field. *Synonym: OPTICAL AXIS. See also: TOLERANCE FIELD.*

fiber bandwidth The lowest frequency at which the magnitude of the fiber transfer function decreases to a specified fraction of the zero frequency value. Often, the specified value is one-half the optical power at zero frequency. *See also: TRANSFER FUNCTION.*

fiber buffer A material that may be used to protect an optical fiber waveguide from physical damage, providing mechanical isolation and/or protection.

Note: Cable fabrication techniques vary, some resulting in firm contact between fiber and protective buffering, others resulting in a loose fit, permitting the fiber to slide in the buffer tube. Multiple buffer layers may be used for added fiber protection. *See also: FIBER BUNDLE.*

fiber bundle   An assembly of unbuffered optical fibers. Usually used as a single transmission channel, as opposed to multifiber cables, which contain optically and mechanically isolated fibers, each of which provides a separate channel.
Note 1. Bundles used only to transmit light, as in optical communications, are flexible and are typically unaligned.
Note 2. Bundles used to transmit optical images may be either flexible or rigid, but must contain aligned fibers. *See also: ALIGNED BUNDLE; FERRULE; FIBER OPTICS; MULTIFIBER CABLE; OPTICAL CABLE; OPTICAL FIBER; PACKING FRACTION.*

fiber cable   *See: MULTIPLE-FIBER CABLE; MULTI-CHANNEL SINGLE-FIBER CABLE; SINGLE-CHANNEL SINGLE-FIBER CABLE.*

fiber cable assembly   *See: MULTIPLE-FIBER CABLE ASSEMBLY.*

fiber cladding   A light-conducting material that surrounds the core of an optical fiber and that has a lower refractive index than the core material.

fiber coating   *See: OPTICAL FIBER COATING.*

fiber core   The central light-conducting portion of an optical fiber.
Note: The core has a higher refractive index than the cladding that surrounds it.

fiber core diameter   In an optical fiber, the diameter of the higher refractive index medium that is the primary transmission medium for the fiber.

fiber coupling   *See: SOURCE-FIBER COUPLING.*

fiber crosstalk   In an optical fiber, exchange of lightwave energy between a core and the cladding, the cladding and the ambient surrounding, or between differently-indexed layers.
Note: Fiber crosstalk is usually undesirable, since differences in path length and propagation time can result in dispersion, reducing transmission distances. Thus, attenuation is deliberately introduced into the cladding by making it lossy.

fiber-detector coupling   In fiber optic transmission systems, the transfer of optical signal power from an optical fiber to a detector for conversion to an electrical signal.
Note: Many optical fiber detectors have an optical fiber pigtail for connection by means of a splice or a connector to a transmission fiber.

fiber diameter   The diameter of an optical fiber, normally inclusive of the core, the cladding if step-indexed, and any adherent coating not normally removed when making a connection, such as by a butted or tangential connection.

fiber dispersion   The lengthening of the width of an electromagnetic-energy pulse as it travels along a fiber, caused by material dispersion, due to the frequency dependence of the refractive index, modal dispersion, caused by different group velocities of the different modes, and waveguide dispersion due to frequency dependence of the propagation constant of that mode.

fiber faceplate   A coherent array of fused optical fibers used as a cover for a light source, such as a LED or a vacuum or gas tube, usually cut from a boule. *See also: BOULE.*

fiber jacket   *See: OPTICAL FIBER JACKET.*

fiber junction   *See: OPTICAL FIBER JUNCTION.*

fiber light-guide   *See: OPTICAL FIBER.*

fiber loss   *See: SOURCE-TO-FIBER LOSS.*

fiber-optic cable   Optical fibers incorporated into an assembly of materials that provides tensile strength, external protection, and handling properties comparable to those of equivalent-diameter coaxial cables.
Note: Fiber-optic cables (light guides) are a direct replacement for conventional coaxial cables and wire pairs. The glass-based transmission facilities occupy far less physical volume for an equivalent transmission capacity, which is a major advantage in crowded underground ducts. In addition it may be that they can be manufactured for far less and that installation and maintenance costs can be less. These advantages, with the reduced use of critical metals, such as copper, is a strong impetus for rapid development of lightwave communication systems.

fiber-optic communications (FOC)   Communication systems and components in which optical fibers are used to carry signals from point to point.

fiber-optic connector *See:* FIXED FI-
BER-OPTIC CONNECTOR; FREE FIBER-
OPTIC CONNECTOR.

fiber-optic multiport coupler An optical unit, such as a scattering or diffusion solid "chamber" of optical material,
that has at least one input and two outputs, or at least two inputs and one output, that can be used to couple various
sources to various receivers.
Note: The ports are usually optical fibers.
If there is only one input and one output
port, it is simply a connector.

fiber-optic probe A flexible probe made
up of a bundle of fine glass fibers optically aligned to transmit an image.

fiber-optic rod coupler A graded-
index cylindrically-shaped section of
optical fiber or rod with a length corresponding to the pitch of the undulations
of lightwaves caused by the graded refractive index, the light beam being injected via fibers at an off-axis end-point
on the radius, with the undulations of
the resulting wave varying periodically
from one point to another along the rod
and with half-reflection layers at the $\frac{1}{4}$
pitch point of the undulations providing
for coupling between input and output
fibers.

fiber-optic rod multiplexer-filter A
graded-index cylindrically-shaped section of optical fiber or rod with a length
corresponding to the pitch of the undulations of lightwaves caused by the
graded refractive index, the light beam
being injected via fibers at an off-axis
end-point on the radius, with the
undulations of the resulting wave varying periodically from one point to another along the rod and with interference
layers at the $\frac{1}{4}$ pitch point of the undulations, providing for multiplexing or
filtering.

fiber-optic scrambler Similar to a fiberscope except that the middle section
of loose fiber is deliberately disoriented
as much as possible. Then potted and
sawed, each half is then capable of coding a picture that can be decoded by the
other half.

fiber optics (FO) The branch of optical
technology concerned with the transmission of radiant power through fibers
made of transparent materials such as
glass, fused silica, or plastic.
Note: 1. Telecommunication applications
of fiber optics employ flexible fibers. Ei-

ther a single discrete fiber or a nonspa-
tially aligned fiber bundle may be used
for each information channel. Such fibers are often referred to as "optical
waveguides" to differentiate from fibers
employed in noncommunications applications.
Note: 2. Various industrial and medical
applications employ (typically high-loss)
flexible fiber bundles in which individual fibers are spatially aligned, permitting optical relay of an image. An example is the endoscope.
Note: 3. Some specialized industrial applications employ rigid (fused) aligned
fiber bundles for image transfer. An example is the fiber optics faceplate used
on some high-speed oscilloscopes.

fiber-optic splice A nonseparable junction joining one optical conductor to another.

fiber-optic terminus A device, used to
terminate an optical conductor, that provides a means to locate and contain an
optical conductor within a connector.

**fiber optic transmission system
(FOTS)** A transmission system utilizing small diameter transparent fibers
through which light is transmitted.
Note: Information is transferred by modulating the transmitted light. These
modulated signals are detected by light-
sensitive devices, i.e., photodetectors. *See
also:* LASER FIBER-OPTIC TRANSMIS-
SION SYSTEM.

fiber-optic waveguide A relatively
long thin strand of transparent substance, usually glass, capable of conducting an electromagnetic wave of optical
wavelength (visible or near-visible region of the frequency spectrum) with
some ability to confine longitudinally
directed, or near-longitudinally-directed,
light waves, to its interior by means of
internal reflection.
Note: The fiber-optic waveguide may be
homogeneous or radially inhomogen-
eous with step or graded changes in its
index of refraction, the indices being
lower at the outer regions, the core thus
being of increased index of refraction.

fiber preform *See:* OPTICAL FIBER PRE-
FORM.

fiber scattering In an optical fiber, the
coupling, or leaking, of lightwave power
out of the core of the fiber by Rayleigh
scattering or guide imperfections such as
dielectric strain, compositional or physi-

59

cal discontinuities in the core or cladding, irregularities and extraneous inclusions in the core-cladding interface, curvature of the optical axis, or tapering. *Note:* Scattering losses are measured in all directions as an integrated effect and expressed in dB/km.

fiberscope A device consisting of an entry point, at which a bundle of optical fibers can enter, and a faceplate surface, on which the entering fibers can uniformly terminate, in order to display the optical image received through the fibers.
Note: The bundle of fibers transmit a full color image that remains undisturbed when the bundle is bent. By mounting an objective lens on one end of the bundle, and an eyepiece at the other, the assembly becomes a flexible fiberscope that can be used to view objects that otherwise would be inaccessible for direct viewing.

fiber transfer function *See: OPTICAL FIBER TRANSFER FUNCTION.*

fiber trap *See: OPTICAL FIBER TRAP.*

field *See: ADDRESS FIELD; ADDRESS FIELD EXTENSION; INFORMATION FIELD; VIEW FIELD.*

field coupling *See: EVANESCENT-FIELD COUPLING.*

field curvature In optics, an abberration of actions that causes a plane image, i.e. a flat image, to be focussed onto a curved surface instead of a flat plane.

field-effect transistor Electronic device used in conjunction with light sources or detectors, as a high-speed driver (source + FET) or a low-noise amplifier (PIN detector + FET). Monolithic transmitters, receivers and repeaters can be realized by the integration of FETs and optoelectronic components.

field lens In an optical system or instrument, a positive lens used to collect the chief rays, i.e. field rays, of image forming bundles so that the entire bundles, or sufficient portions of them, will pass through the exit pupil of the instrument. *Note:* A field lens is usually located at or near the focal point of the objective lens. The field lens increases the size of the field that can be viewed with any given eyelens diameter.

field rays In the object space of a symmetrical optical system, a ray that intersects the optical axis at the center of the entrance pupil of the optical system. *Note:* In the image space, the same ray

emerges from the exit pupil. In a thick lens, a field ray is a principal ray.

field strength The intensity of an electric, magnetic, or electromagnetic field at a given point. *Note:* Normally used to refer to the rms value of the electric field, expressed in volts per meter, or of the magnetic field, expressed in amperes per meter. *Synonym: RADIO FIELD INTENSITY.*

film *See: MULTILAYER DIELECTRIC FILM; PHOTOCONDUCTIVE FILM.*

film optical modulator *See: THIN-FILM OPTICAL MODULATOR.*

film optical multiplexers *See: THIN-FILM OPTICAL MULTIPLEXERS.*

film optical switch *See: THIN-FILM OPTICAL SWITCH.*

film optical waveguide *See: THIN-FILM OPTICAL WAVEGUIDE.*

filter 1. In an optical system, a device with the desired characteristics of selective transmittance and optical homogeneity used to modify the spectral composition of radiant light flux (A filter is usually made of special glass, gelatin, or plastic optical parts with plane parallel surfaces that are placed in the path of light through the optical system of an instrument to selectively absorb certain wavelengths of light, reduce glare, or reduce light intensity. Colored, ultraviolet, neutral density, and polarizing filters are in common use. Filters may be separate elements or integral devices mounted so that they can be placed in or out of position in a system as desired.). 2. A network that passes desired frequencies but greatly attenuates other frequencies. 3. A device for use on power lines or signal lines, specifically designed to pass only selected frequencies and to attenuate substantially all other frequencies. There are two basic types of filters: (a) active filters: those which require the application of power and (b) passive filters: those which use inductance-capacitance components and do not require the application of power for the utilization of their filtering properties. *See also: BAND-STOP FILTER; BANDPASS FILTER; DICHROIC FILTER; HIGH-PASS FILTER; LINE FILTER BALANCE; LOW-PASS FILTER; FIBER-OPTIC ROD MULTIPLEXER-FILTER; OPTICAL FILTER; ROOFING FILTER.*

filter-coupler-switch-modulator *See: INTEGRATED-OPTICAL CIRCUIT FILTER-COUPLER-SWITCH-MODULATOR.*

finish   *See: OPTICAL END-FINISH.*

finished lens   A lens having both surfaces ground and polished to specific dioptric power or focus.

five-layer four-heterojunction diode   A four-heterojunction laser diode, consisting of two pairs of heterojunctions, that has five layers of step-indexed material, i.e. five layers of material with a sudden transition of refractive index at the interface between layers, so as to confine the emitted light to a narrow beam for optimum coupling to an optical fiber, fiber bundle, or integrated optical circuit.

Note: Usually only three different refractive indices are involved, since there may be a pair of identically-indexed outside layers, a pair of identically-indexed inside layers on opposite sides of a center layer, each pair and center layer being of different refractive index, with decreased refractive indices toward the outside, resulting in a layered cross-section with step-indices of $N(1) : N(2) : N(3) : N(2) \, N(1)$, with $N(1)$ less than $N(3)$. Thus, almost all of the generated and emitted light is confined to the center layer by internal reflection.

fixed fiber-optic connector   A connector that permits connection of optical fiber components that are to be associated on a permanent basis.

Note: Fixed fiber-optic connectors can be used to connect source to optical conductor, optical conductor to optical conductor, or optical conductor to detector. They are usually part of the devices being connected.

fixed focus   Pertaining to instruments that are not provided with a means of focusing.

fixed optical attenuator   A device that attenuates the intensity of light waves, when inserted into an optical waveguide link, a fixed or given number of dB, for example, a standard fixed single attenuation of 3, 6, 10, 20, or 40 dB for each attenuator.

fixed-reference modulation   A type of modulation in which the choice of the significant condition for any signal element is based on a fixed reference.

flat fading.   Fading in which all frequency components of the received radio signal vary in the same proportion simultaneously.

flat weighting   In a noise measuring set, an amplitude-frequency characteristic which is flat over a specified frequency range, which must be stated. Flat noise power may be expressed in $dBrn(f_1\text{-}f_2)$, or in $dBm(f_1\text{-}f_2)$. The terms 3 kHz flat weighting and 15 kHz flat weighting are also used for characteristics which are flat from 30 Hz to the upper frequency indicated. *See also: dBrn(f_1-f_2).*

flip chip   In fiber optic circuits and Integrated Optical Circuits (IOC) an optical switch designed to control light conduction paths into and out of a junction. *See also: OPTICAL SWITCH.*

flutter   1. The distortion due to variation in loss resulting from the simultaneous transmission of a signal at another frequency. 2. The distortion due to variation in loss resulting from phase distortion. 3. In recording and reproducing, deviation of frequency which results, in general, from irregular motion during recording, duplication, or reproduction. 4. In radio transmission, rapidly changing signal amplitude levels together with variable multipath time delays, caused by the reflection and possible partial absorption of the radio signal from aircraft flying through the beam or common scatter volume. 5. The effect of the variation in the transmission characteristics of a loaded telephone circuit caused by the action of telegraph direct currents on the loading coils. 6. Fast changing variation in received signal strength, such as may be caused by atmospheric variations, antenna movements in a high wind, or interaction with another signal. *See also: DISTORTION.*

flux   Obsolete synonym for radiant power.

FM   *Acronym for FREQUENCY MODULATION.*

FM improvement factor   The signal-to-noise ratio at the output of the receiver divided by the carrier-to-noise ratio at the input of the receiver. This improvement is always obtained at the price of an increased bandwidth in the receiver and the transmission path.

FM improvement threshold   The point in an FM receiver at which the peaks in the rf signal equal the peaks of the thermal noise generated in the receiver.

Note: A baseband signal-to-noise ratio of about 30 dB is typical at the improvement threshold, and this ratio improves 1 dB for each dB of increase in the signal above the threshold.

FO *Acronym for FIBER OPTICS.*

FOC *Acronym for FIBER-OPTIC COMMU-NICATIONS.*

focal length The distance from a lens, or some point therein, or from a mirror, to the image of a small, infinitely distant source of light.
Note: This image point is referred to as the focal point. *See also: BACK FO-CAL LENGTH: EQUIVALENT FOCAL LENGTH; FRONT FOCAL LENGTH.*

focal plane In an optical system, a plane through the focal point perpendicular to the principal axis of the system, such as a lens or mirror, for example, the film plane in a camera focused at infinity.

focal point 1. In an optical system, the point at which a bundle of rays form a sharp image of an object. 2. The point at which an object in an optical system must be placed for a sharp image to be obtained. *Synonym: PRINCIPAL FOCUS.*

focus In an optical system, to adjust the system, such as the eyepiece or objective of a microscope, telescope, or camera, so that the image is clearly seen by the observer or so that a sharp, distinct image is registered. *See also: FIXED FOCUS; FOCAL POINT.*

focus point *See: PRINCIPAL FOCUS POINT.*

foot-candle A unit of illuminance equal to one lumen incident per square foot.
Note: It is the illuminance of a surface placed one foot from a light source having a luminous intensity of one candle or candela.

footprint That portion of the earth's surface illuminated by a narrow beam from a satellite. It is less than earth coverage. *See also: EARTH COVERAGE; SATELLITE.*

forward channel The channel of a data circuit that transmits data from the data source to the data sink. *See also: BACK-WARD CHANNEL; DATA TRANSMISSION.*

forward error correction (FEC) A system of error control for data transmission wherein the receiving device has the capability to detect and correct any character or code block which contains a predetermined number of bits in error.
Note: FEC is accomplished by adding bits to each transmitted character or code block using a predetermined algorithm. *See also: ERROR.*

forward propagation ionospheric

scatter (FPIS) *Synonym: IONOSPHER-IC SCATTER.*

forward scatter 1. The deflection by reflection or refraction of an electromagnetic wave or signal in such a manner that a component of the wave is deflected in the direction of propagation of the incident wave or signal. 2. The component of an electromagnetic wave or signal that is deflected by reflection or refraction in the direction of propagation of the incident wave or signal. 3. To deflect, by reflection or refraction, an electromagnetic wave or signal in such a manner that a component of the wave or signal is deflected in the direction of propagation of the incident wave or signal.
Note: The term scatter can be applied to reflection or refraction by relatively uniform media but it is usually taken to mean propagation in which the wavefront and direction are modified in a relatively disorderly fashion. *See also: BACKSCATTER; PROPAGATION.*

forward signal A signal sent in the direction from the calling to the called station, or from a data source to a data sink.
Note: The forward signal is transmitted in the forward channel.

forward voltage The positive voltage that has to be applied to the p-side of an optoelectronic device at a typical operating point. Mostly a measure of the contact quality. Symbol: V_F. Typical AlGaAs diode-laser V_F values are 1.6–1.8 volts.

FOTS *Acronym for FIBER OPTICS TRANS-MISSION SYSTEM.*

Foucault knife-edge test A method of determining the errors in an image of a point source by partially occluding the light from an image by means of a knife edge. The same test may be used to measure the errors in refracting or reflecting surfaces.

four-heterojunction diode A laser diode with two double heterojunctions, i.e. two pairs of heterojunctions to provide improved control of direction of radiation and radiative recombination. *Synonym: SYMMETRICAL DOU-BLE-HETEROJUNCTION DIODE. See also: FIVE-LAYER FOUR-HETEROJUNCTION DIODE.*

FPIS *Acronym for FORWARD PROPAGA-TION IONOSPHERIC SCATTER.*

fraction *See: PACKING FRACTION.*

fraction loss *See: PACKING FRACTION LOSS.*

frame 1. In data transmission, the sequence of contiguous bits bracketed by and including beginning and ending flag sequences.

Note: A typical frame might consist of a specified number of bits between flags and contain an address field, a control field, and a frame check sequence. A frame may or may not include an information field.

2. In the multiplex structure of PCM systems, a set of consecutive digit time slots in which the position of each digit time slot can be identified by reference to a frame alignment signal.

Note: The frame alignment signal does not necessarily occur, in whole or in part, in each frame.

3. In a TDM system, a repetitive group of signals resulting from a single sampling of all channels, including any additional signals for synchronizing and other required system information.

Note: In-frame is the condition which exists when there is a channel-to-channel and bit-to-bit correspondence (exclusive of transmission errors) between all inputs of a time division multiplexer and output of its associated demultiplexer.

4. In facsimile systems, a rectangular area, the width of which is the available line and the length of which is determined by the service requirements. *See also: DATA TRANSMISSION; FLAG SEQUENCE; INTERFRAME TIME FILL; OUT-OF-FRAME-ALIGNMENT TIME; PULSE-CODE MODULATION; REFRAMING TIME; RESPONSE FRAME.*

Fraunhofer diffraction pattern *Synonym: FAR-FIELD DIFFRACTION PATTERN.*

free-carrier absorption loss Internal cavity loss in diode lasers due to light absorption by carriers injected in the active layer or by the dopant in the cladding layer(s).

free fiber-optic connector A connector that permits connection of optical fiber components that also permits easy disconnection.

Note: Free fiber-optic connectors can be used to connect source to optical conductor, optical conductor to optical conductor, or optical conductor to detector. They may be cable mounted, but are independent of components.

free space A theoretical concept of space devoid of all matter.

Note: The term also implies remoteness from material objects that could influence the propagation of electromagnetic waves.

free space loss The signal attenuation that would result if all obstructing, scattering, or reflecting influences were sufficiently removed so as to have no effect on propagation.

Note: Free space loss is primarily caused by beam divergence, i.e., signal energy spreading over larger areas at increased distances from the source.

frequency The number of cycles or events per unit of time. When the unit of time is one second, the measurement unit is the hertz (Hz). *See also: ASSIGNED FREQUENCY; AUDIO FREQUENCIES; AUTHORIZED FREQUENCY; BAND; BANDWIDTH; CARRIER FREQUENCY; CENTER FREQUENCY; CHARACTERISTIC FREQUENCY; CRITICAL FREQUENCY; CUT-OFF FREQUENCY; FRAME FREQUENCY; INSERTION LOSS VS FREQUENCY CHARACTERISTIC; LOWEST USEFUL HIGH FREQUENCY; MAXIMUM KEYING FREQUENCY; MAXIMUM MODULATING FREQUENCY; MAXIMUM USABLE FREQUENCY; MEDIUM FREQUENCY; OPTIMUM TRAFFIC FREQUENCY; PRECISE FREQUENCY; RADIO FREQUENCY; REFERENCE FREQUENCY; SAMPLING FREQUENCY; TRANSITION FREQUENCY; SIGNAL FREQUENCY SHIFT; SINGLE FREQUENCY INTERFERENCE; SPECTRUM DESIGNATION OF FREQUENCY; VOICE FREQUENCY.*

frequency averaging A process by which network synchronization is achieved by use of oscillators at all nodes which adjust their frequencies to the average frequency of the digital bit streams received from connected nodes. All oscillators are assigned equal weight in determining the ultimate network frequency since there is no reference oscillator. *See also: DEMOCRATICALLY SYNCHRONIZED NETWORK.*

frequency-change signaling A signaling method in which one or more particular frequencies correspond to each desired signaling condition of a code. The transition from one set of frequencies to the other may be either a continuous or a discontinuous change in frequency or in phase.

frequency deviation In frequency modulation, the peak difference between the instantaneous frequency of the modulated wave and the carrier frequency.

frequency displacement The end-to-end shift in frequency that may result from independent frequency translation errors in a circuit. *See also: FREQUENCY TRANSLATION.*

frequency distortion *See: AMPLITUDE-VERSUS-FREQUENCY DISTORTION.*

frequency diversity Any method of diversity transmission and reception wherein the same information signal is transmitted and received simultaneously on two or more independently fading carrier frequencies. *See also: DIVERSITY RECEPTION; DUAL DIVERSITY.*

frequency division multiple access In satellite communications, the use of frequency division to provide multiple and simultaneous transmissions to a single transponder.

frequency division multiplex (FDM) A method of deriving two or more simultaneous, continuous channels from a transmission medium connecting two points by assigning separate portions of the available frequency spectrum to each of the individual channels. *Synonym: CARRIER MULTIPLEX. See also: MULTIPLEX; WAVELENGTH DIVISION MULTIPLEX.*

frequency drift A slow, undesired change in the frequency of an oscillator, transmitter, receiver, or other equipment.

frequency exchange signaling A frequency change signaling method in which the change from one signaling condition to another is accompanied by decay in amplitude of one or more frequencies and by buildup in amplitude of one or more other frequencies. *Synonym: TWO-SOURCE FREQUENCY KEYING.*

frequency frogging 1. The interchanging of the frequency allocations of carrier channels to prevent singing, reduce crosstalk, and to correct for line slope (This can be accomplished by having the modulators in a repeater translate a low-frequency group to a high-frequency group, and vice versa. Because of this frequency inversion process, a channel will appear in the low group for one

repeater section and will then be translated to the high group for the next section. This results in nearly constant attenuation with frequency over two successive repeater sections, and eliminates the need for large slope equalization and adjustment. Also, singing and crosstalk are minimized because the high level output of a repeater is at a different frequency from the low level input to other repeaters.). 2. Alternate use of two frequencies at repeater sites of line-of-sight microwave systems.

frequency guard band A frequency band left vacant between two channels to provide a margin of safety against mutual interference. *See also: TIME GUARD BAND.*

frequency modulation (FM) The form of angle modulation in which the instantaneous frequency of a sine wave carrier is caused to depart from the carrier frequency by an amount proportional to the instantaneous value of the modulating signal.
Note: Combinations of phase and frequency modulation are also commonly referred to as frequency modulation. *See also: PHASE MODULATION.*

frequency offset The fractional frequency deviation of a frequency with respect to another frequency, viz:

$$\Delta f/f = (f_1 - f_2)/f_2$$

where Δf is the difference between the two frequencies, f_1 and f_2, and f is the reference frequency, f_2, with respect to which the offset is taken.
Note: An example is the difference between the frequency of the National Bureau of Standards Primary Frequency Standard and the quart-crystal controlled oscillators from which the NBS broadcast signals are derived. These offsets may be on the order of one part in 10^{12}.

frequency response *Synonym: TRANSFER FUNCTION (of a device).*

frequency shift 1. A type of telegraph operation in which the mark and space signals are different frequencies. 2. Any change in the frequency of a radio transmitter or oscillator. (Also called rf shift.) 3. In facsimile, a frequency modulation system where one frequency represents picture black and another frequency represents picture white. Frequencies between these two limits may represent shades of gray. 4. The number of hertz difference in a frequency modulation system.

frequency shift keying (FSK) A form of frequency modulation in which the modulating signal shifts the output frequency between predetermined values and in which the output signal has no phase discontinuity.
Note: Commonly, the instantaneous frequency is shifted between two discrete values termed the mark and space frequencies. *Synonyms: FREQUENCY SHIFT MODULATION; FREQUENCY SHIFT SIGNALING. See also: TWO-TONE KEYING.*

frequency shift modulation *Synonym: FREQUENCY SHIFT KEYING.*

frequency shift signaling *Synonym: FREQUENCY SHIFT KEYING.*

frequency translation The transfer en bloc of signals occupying a definite frequency band (such as a channel or group of channels) from one position in the frequency spectrum to another, in such a way that the arithmetic frequency difference of signals within the band is unaltered. *See also: FREQUENCY DISPLACEMENT.*

Fresnel diffraction pattern *Synonym: NEAR-FIELD DIFFRACTION PATTERN.*

Fresnel reflectance loss *See: FRESNEL REFLECTION LOSS.*

Fresnel reflection The reflection of a portion of the light incident on a planar interface between two homogeneous media having different refractive indices.
Note: 1. Fresnel reflection occurs at the air-glass interfaces at entrance and exit ends of an optical waveguide. Resultant transmission losses (on the order of 4% per interface) can be virtually eliminated by use of antireflection coatings or index matching materials.
Note: 2. Fresnel reflection depends upon the index difference and the angle of incidence; it is zero at Brewster's angle for one polarization. In optical elements, a thin transparent film is sometimes used to give an additional Fresnel reflection that cancels the original one by interference. This is called an antireflection coating. *See also: ANTIREFLECTION COATING; BREWSTER'S ANGLE; INDEX MATCHING MATERIAL; REFLECTANCE; REFLECTION; REFRACTIVE INDEX.*

Fresnel reflection loss The power loss

incurred at an interface surface when an electromagnetic wave is incident upon it and part of the incident power is reflected.
Note: The reflection loss depends on many factors, including the indices of refraction of the incident and refracting media, the wavelength, the angle of incidence, and the incident light polarization relative to the interface.
Note: Reflection losses that are incurred at the input and output of an optical fiber are due to the difference in refractive index between the fiber and the medium from which the light enters and to which it leaves. *Synonym: FRESNEL REFLECTANCE LOSS. See also: REFLECTION COEFFICIENT; TRANSMISSION COEFFICIENT.*

Fresnel reflection method The method for measuring the index profile of an optical fiber by measuring the reflectance as a function of position on the end face. *See also: FRESNEL REFLECTION; INDEX PROFILE; REFLECTANCE.*

Fresnel zone A cigar-shaped shell of circular cross section surrounding the direct path between a transmitter and a receiver. For the first Fresnel zone, the distance from the transmitter to any point on this shell and on to the receiver is one half-wavelength longer than the direct path; for the second Fresnel zone, two half-wavelengths, etc. *See also: EFFECTIVE EARTH RADIUS; K-FACTOR; PATH CLEARANCE; PATH PROFILE; PATH SURVEY; PROPAGATION PATH OBSTRUCTION.*

fringe In optics, a light or dark band caused by interference of two or more electromagnetic waves, usually light waves, so that areas (bands) of reinforcement and cancellation occur. *See also: NEWTON'S FRINGES.*

front-emitting LED *See: SURFACE-EMITTING LED.*

front-end noise temperature A measure of the thermal noise in the first stage of a receiver. *See also: NOISE.*

front focal length (FFL) In an optical system, the distance measured from the principal focus located in the front space, to the first principal point.

front-surface mirror An optical mirror on which the reflecting surface is applied to the front surface of the mirror instead of the back, i.e. to the surface of first incidence.

Note: The reflected light does not pass through any substrate. *See also: BACK SURFACE MIRROR.*

front-to-back ratio A ratio of parameters used in connection with antennas, rectifiers, or other devices in which signal strength or resistance, or other parameters, in one direction is compared with that in the opposite direction.

FSK *Acronym for FREQUENCY SHIFT KEYING.*

full-duplex operation *Synonym: DUPLEX OPERATION.*

full-width at half-power Parameter characterizing the output beams of diode lasers or LEDs. The width of the beam at half intensity in a given plane. Symbol: FWHP. Example: For a diode laser the beam is usually characterized by FWHP values in planes parallel and perpendicular to the junction plane, and intersecting the source.

full width (duration) half maximum A measure of the extent of a function. Given by the difference between the two extreme values of the independent variable at which the dependent variable is equal to half of its maximum value. The term "duration" is preferred when the independent variable is time.

Note: Commonly applied to the duration of pulse waveforms, the spectral extent of emission or absorption lines, and the angular or spatial extent of radiation patterns.

function *See: MODULATION TRANSFER FUNCTION; OPTICAL FIBER TRANSFER FUNCTION.*

fundamental mode The lowest order mode of a waveguide. In fibers, the mode designated LP_{01} or HE_{11}. *See also: MODE.*

furcation coupling The mixing of signals from several separate optical fibers by passing them through a common single fiber rod thus obtaining a signal containing all the components of the several signals.

Note: The mixing of several colors can take place in this manner.

fused quartz Glass made by melting natural quartz crystals; not as pure as vitreous silica.

fused silica *Synonym: VITREOUS SILICA. See also: FUSED QUARTZ.*

fusion splice A splice accomplished by the application of localized heat sufficient to fuse or melt the ends of two lengths of optical fiber, forming a continuous, single fiber.

FWHM *Acronym for* full width at half maximum. *See also: FULL WIDTH (DURATION) HALF MAXIMUM.*

FWHP *Acronym for FULL-WIDTH HALF-POWER.*

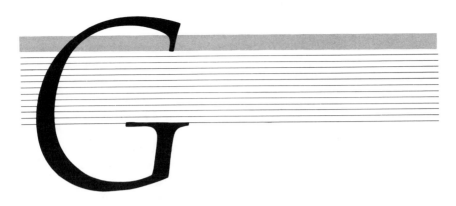

GaAs *Chemical symbol for GALLIUM AR-SENIDE.*

gain The ratio of output current, voltage, or power to input current, voltage or power, respectively.

Note: 1. Gain is usually expressed in dB.

Note: 2. Differences in power levels between points in a system may be expressed as gains. *See also: ANTENNA GAIN-TO-NOISE TEMPERATURE; DIRECTIVE GAIN; EQUAL GAIN COMBINER; HEIGHT GAIN; INSERTION GAIN; LOOP GAIN; NET GAIN; POWER GAIN OF AN ANTENNA.*

gain-bandwidth product The product of the gain of an active device and a specified bandwidth.

Note: for an avalanche photodiode. The gain-bandwidth product is the gain times the frequency of measurement when the device is biased for maximum obtainable gain.

gain coefficients Coefficients in the relationship between the optical gain and the injected current density, characterizing the active-layer material of a semiconductor laser.

Note: See for example F. Stern, *J. Appl. Phys.,* vol. 47, pp. 5382–5386, Dec. 1976.

gain factor *See: PHOTOCONDUCTIVE GAIN FACTOR*

Gain-guided diode laser A diode laser for which lateral mode control is achieved only by the current distribution in the plane of the junction (i.e., gain guiding). Gain-guided devices have multi-longitudinal-mode spectra and thus are sought for graded-index multimode-fiber systems, to reduce modal noise. Gain-guided lasers have highly astigmatic beams.

gain guiding In planar geometry DH lasers the mechanism by which the injected carriers confine the optical mode in the plane of the junction. The effects of a lateral variation in carrier profile are to create both a negative-index contribution (antiguiding) due to the local increase in carrier concentration as well as a positive-index contribution in the imaginary part of the dielectric constant due to the local increase in optical gain.

Note: See for example P. M. Asbeck, *et al., IEEE J. Quantum Electron.,* vol. 15, pp. 727–734, 1979.

gain-induced guiding The contribution to the dielectric constant of variations in optical gain. The part of the gain guiding mechanism that corresponds to variations in the imaginary part of the dielectric constant. *See: GAIN GUIDING.*

galactic radio noise *Synonym for* COSMIC NOISE.

gallium antimonide A binary semiconductor compound used as substrate and/or active layer for lattice-matched AlGaAsSb/GaSb structures. Used mostly as substrate for APDs. Bandgap wavelength: 1.8 μm. *See also ALUMINUM GALLIUM ANTIMONIDE AND GALLIUM ARSENIDE ANTIMONIDE.*

gallium arsenide A binary semiconductor compound used as substrate and/or active layer for lattice-matched AlGaAs/GaAs and InGaAsP/GaAs light-emitting structures. Used mostly for AlGaAs lasers and LEDs emitting in the 0.7–0.9 μm wavelength range. The InGaAsP/GaAs structures can emit mostly in the visible. Bandgap wavelength: 0.87 μm. *See also: ALUMINUM GALLIUM ARSENIDE.*

gap loss That optical power loss caused by a space between axially aligned fibers.

Note: For waveguide-to-waveguide coupling, it is commonly called "longitudinal offset loss." *See also: COUPLING LOSS.*

garble An error in transmission, reception, encryption, or decryption which changes the text of a message or any portion thereof in such a manner that it is incorrect or undecryptable.

garnet source *See: YAG/LED SOURCE.*

GaSb *Chemical symbol for GALLIUM ANTIMONIDE.*

gas laser A laser in which the active medium is a gas.

Note: Types of lasers include the atomic laser, such as the helium-neon laser; the ionic laser, such as the argon, krypton, and xenon lasers; the metal-vapor laser, such as the helium-cadmium and helium-selenium lasers; and the molecular laser, such as the carbon-dioxide, hydrogen-cyanide and water-vapor lasers. *See also: MIXED-GAS LASER.*

gate 1. A device having one output channel and one or more input channels, such that the output channel state is completely determined by the input channel states, except during switching transients. 2. A combinational logic element having at least one input channel.

gating 1. The process of selecting only those portions of a wave between specified time intervals or between specified amplitude limits. 2. The controlling of signals by means of combinational logic elements. 3. A process in which a predetermined set of conditions, which, when established, permits a second process to occur.

Gaussian beam A beam of light whose electric field amplitude distribution is Gaussian. When such a beam is circular in cross section, the amplitude is

$$E(r) = E(0) \exp\left[-(r/w)^2\right]$$

where r is the distance from beam center and w is the radius at which the amplitude is 1/e of its value on the axis; w is called the beamwidth. *See also: BEAM DIAMETER.*

Gaussian pulse A pulse that has the waveform of a Gaussian distribution. In the time domain, the waveform is

$$f(t) = A \exp\left[-(t/a)^2\right]$$

where A is a constant, and a is the pulse half duration at the 1/e points. *See also: FULL WIDTH (DURATION) HALF MAXIMUM.*

geometric image Pertaining to the location and shape of the image of a particle, as predicted by geometric optics alone.

Note: The geometric image is to be distinguished from the diffraction image, determined from considerations of both physical and geometrical optics. With completely corrected objectives, the geometrical image of two points is again two points, but the actual image or the diffraction image may or may not suggest the presence of an object comprised of two points or two small particles.

geometric optics The treatment of propagation of light as rays.

Note: Rays are bent at the interface between two dissimilar media or may be curved in a medium in which refractive index is a function of position. *See also: AXIAL RAY; MERIDIONAL RAY; OPTICAL AXIS; PARAXIAL RAY; PHYSICAL OPTICS; SKEW RAY.*

ghost image A spurious single or multiple image of objects seen in optical instruments. Caused by reflections from optical surfaces.

Note: By coating optical surfaces with low reflection films, the ghost images are greatly reduced.

glass *See: MAGNIFIER.*

glass laser A solid-state laser whose active laser medium is glass.

g-line *See: GOUBAU LINE.*

Goubau line A singlewire open waveguide.

Note: The Goubau line is capable of guiding an axial cylindrical surface wave. Synonym: G-line.

graded fiber *See: DOPED-SILICA GRADED FIBER.*

graded-index fiber An optical fiber with a variable refractive index that is a function of the radial distance from the fiber axis, the refractive index getting progressively lower away from the axis.

Note: This characteristic causes the light rays to be continually refocused by refraction into the core. As a result, there is a designed continuous change in refractive index between the core and cladding along a fiber diameter. *See also: UNIFORM INDEX PROFILE.*

graded index optical waveguide A waveguide having a graded index profile in the core. *See also: GRADED INDEX PROFILE; STEP INDEX OPTICAL WAVEGUIDE.*

graded index profile Any refractive index profile that varies with radius in the core. Distinguished from a step index profile. *See also: DISPERSION; MODE VOLUME; MULTIMODE OPTICAL WAVEGUIDE; NORMALIZED FREQUENCY; OPTICAL WAVEGUIDE; PARABOLIC PROFILE; PROFILE DISPER-*

SION; PROFILE PARAMETER; REFRAC-
TIVE INDEX; STEP INDEX PROFILE;
POWER-LAW INDEX PROFILE.

gradual degradation In a light-emitting diode (LED), a reduction in the ex-
ternally measured quantum efficiency.
Note: In a laser diode, the threshold current density increases and the resulting
incremental quantum efficiency decreases, resulting in reduced power output for given current density without
evidence of facet damage. However, the
power output level can usually be re-
stored by an increase in the current
density. *See also: CATASTROPHIC DEG-
RADATION.*

grating *See: DIFFRACTION GRATING.*

grating chromatic resolving power
The resolving power that determines the
minimum wavelength difference for any
spectral order that can be distinguished
as separate.
Note: The chromatic resolving power for
diffraction gratings is usually stated for
cases in which parallel rays of light are
incident upon the grating and is numerically equal to the number of lines or
ruled spacings per unit distance in the
grating. *See also: DIFFRACTION GRAT-
ING SPECTRAL ORDER.*

grating spectral order *See: DIFFRAC-
TION GRATING SPECTRAL ORDER.*

grazing emergence In optics, a condition in which an emergent ray makes an
angle of 90 degrees to the normal of the
emergent surface of a medium. *See also:
EMERGENCE; NORMAL EMERGENCE.*

grazing incidence Pertaining to light
rays incident at 90-degrees to the normal
to the incident surface. *See also: NOR-
MAL INCIDENCE.*

ground state The lowest orbital kinetic
and potential energy state that an electron of a given element can have. *See also:
EXCITED STATE:*
Note: An electron absorbs a quantum of
energy when it moves from the ground
state to an excited state.

group 1. In frequency division multiplexing, a number of voice channels, either within a supergroup or separately,
which in wideband systems is normally
composed of up to 12 voice channels
occupying the frequency band 60 kHz to
108 kHz. This is CCITT Basic group B. 2.
A supergroup is normally 60 voice channels, or 5 groups of 12 voice channels
each, occupying the frequency band 312
kHz to 552 kHz. 3. A mastergroup is
composed of 10 supergroups or 600 voice
channels.
Note: 1. Basic group A (carrier telephone
systems) is an assembly of 12 channels,
occupying upper sidebands in the 12
kHz to 60 kHz band. It is no longer
mentioned in CCITT Recommendations.
Note: 2. The CCITT standard mastergroup contains 5 supergroups; U.S. commercial carriers use 10 supergroups.
Note: 3. The terms "supermastergroup"
or "jumbo group" are sometimes used
for 6 mastergroups. *See also: HIGH-US-
AGE TRUNK GROUP; PHANTOM
GROUP; THROUGH GROUP;
THROUGH-GROUP EQUIPMENT;
THROUGH SUPERGROUP; THROUGH-
SUPERGROUP EQUIPMENT; TRUNK
GROUP; WIDEBAND SYSTEM.*

group-delay spread *See: MULTIMODE
GROUP-DELAY SPREAD.*

group delay time The rate of change of
the total phase shift with angular frequency through a device or transmission
facility.
Note: Group delay time is the time interval required for the crest of a group of
waves to travel through a device or
transmission facility where the component wave trains have slightly different
individual frequencies. *See also: DELAY
DISTORTION; ENVELOPE DELAY DIS-
TORTION.*

group index (denoted N) For a given
mode propagating in a medium of refractive index n, the velocity of light in
vacuum, c, divided by the group velocity
of the mode. For a plane wave of wavelength λ, it is related thus to the refractive index:

$$N = n - \lambda \, (dn/d\lambda)$$

*See also: GROUP VELOCITY; MATERIAL
DISPERSION PARAMETER.*

group velocity 1. For a particular mode,
the reciprocal of the rate of change of the
phase constant with respect to angular
frequency.
Note: The group velocity equals the
phase velocity if the phase constant is a
linear function of the angular frequency.
2. Velocity of the signal modulating a
propagating electromagnetic wave. *See
also: DIFFERENTIAL MODE DELAY;
GROUP INDEX; PHASE VELOCITY.*

G/T *Acronym for ANTENNA GAIN-TO-
NOISE TEMPERATURE*

guide *See: LIGHT GUIDE; ULTRAVIOLET
LIGHT GUIDE*

guided mode *Synonym: BOUND MODE.*

guided ray In an optical waveguide, a ray that is completely confined to the core. Specifically, a ray at radial position r having direction such that

$$0 \leq \sin \theta(r) \leq [n^2(r) - n^2(a)]^{1/2}$$

where $\theta(r)$ is the angle the ray makes with the waveguide axis, $n(r)$ is the refractive index, and $n(a)$ is the refractive index at the core radius. Guided rays correspond to bound (or guided) modes in the terminology of mode descriptors. *Synonyms: BOUND RAY; TRAPPED RAY. See also: BOUND MODE; LEAKY RAY.*

guide layer The portion of the large-optical-cavity (LOC) structure in which a large part of the optical mode propagates while obtaining gain from the active layer. *See: LARGE OPTICAL CAVITY DIODE.*

Note: See for example D. Botez, *Appl. Phys. Lett.,* vol. 36, pp 190–192, Feb. 1980.

HA1-receiver weighting A noise weighting used in a noise measuring set to measure noise across the HA1-receiver of a 302-type or similar instrument.
Note: 1. The meter scale readings are in dBa(HA1).
Note: 2. HA1 noise weighting is obsolete for DoD applications.

half-duplex circuit A circuit that affords communication in either direction but only in one direction at a time. *See: also: DUPLEX CIRCUIT; PUSH-TO-TALK OPERATION; PUSH-TO-TYPE OPERATION.*

half-duplex operation 1. That type of simplex operation which uses a half-duplex circuit. 2. Pertaining to an alternate, one way at a time, transmission mode of operation. *Synonyms: ONE-WAY REVERSIBLE OPERATION; TWO-WAY ALTERNATE OPERATION. See also: DUPLEX OPERATION; SIMPLEX OPERATION.*

half-power points *See: EMISSION-BEAM-ANGLE BETWEEN-HALF-POWER-POINTS*

hamming distance *Synonym: SIGNAL DISTANCE.*

harmonic distortion The presence of frequencies at the output of a device, caused by nonlinearities within the device, which are harmonically related to a single frequency applied to the input of the device. The frequency of the first harmonic at the output (also known as the fundamental frequency) is the input frequency. *See also: SINGLE-HARMONIC DISTORTION; TOTAL HARMONIC DISTORTION.*

harness *See: OPTICAL HARNESS.*

harness assembly *See: OPTICAL HARNESS ASSEMBLY.*

harness run *See: CABLE RUN.*

HDLC *Acronym for HIGH-LEVEL DATA LINK CONTROL.*

head *See: LASER HEAD.*

heatsink High-thermal-conductivity mount that dissipates the heat generated during cw or near-cw operation of diode lasers, LEDs, or microwave devices. Typical heatsink materials are copper and diamond.

Heaviside layer *See: IONOSPHERE.*

heavy seeding In an optical medium, such as glass, pertaining to a condition in which the fine and coarse seeds are very numerous, such as 25 or more to the square inch.

height gain For a given propagation mode of an electromagnetic wave, the ratio of the field strength at a specified height to that at the earth's surface.

HE$_{11}$ mode Designation for the fundamental mode of an optical fiber. *See FUNDAMENTAL MODE.*

Hermite-Gaussian modes Resonant modes of stripe-geometry diode lasers with weak perpendicular and/or lateral confinement. These modes can be expressed in terms of Hermite-Gaussian functions, which occur as solutions of Maxwell's equations in a dielectric medium of dielectric constant that varies parabolically in space.
Note: For examples see Kressel and Butler, *Semiconductor Lasers and Heterojunction LEDs,* Academic Press, New York, 1977.

heterochronous A relationship between two signals such that their corresponding significant instants do not necessarily occur at the same time.
Note: Two signals having different nominal signaling rates, and not stemming from the same clock or from homochronous clocks are usually heterochronous. *See also: ANISOCHRONOUS; HOMOCHRONOUS; ISOCHRONOUS; MESOCHRONOUS; PLESIOCHRONOUS.*

heterodyne repeater A repeater for a radio system in which the received signals are converted to an intermediate frequency, amplified, and reconverted to a new frequency band for transmission over the next repeater section. It is sometimes called an IF repeater.

heteroepitaxial optical waveguide An optical-wavelength electromagnetic waveguide consisting of an optical quality crystal substrate upon which are deposited one or more layers of substances with different indices of refraction, such as cubic heteroepitaxial films of zinc sulphide (ZnS) on gallium arsenide (GaAs) substrate, and zinc selenide (ZnSe) on gallium arsenide (GaAs), with closely matched lattice structures and indices of refraction less than that of the substrate, so that the films themselves do not act as ordinary waveguides with total internal reflection. But optical propagation of leaky modes does occur, with attenuation losses directly proportional to the square of the wavelength.

heterogeneous multiplexing Multiplexing in which all the information bearer channels do not operate at the same data signaling rate. *See also: MULTIPLEX.*

heterojunction A junction between semiconductors that differ in their atomic or alloy compositions. *See also: HOMOJUNCTION.*

heterojunction diode *See: DOUBLE HETEROJUNCTION DIODE; FIVE-LAYER FOUR-HETEROJUNCTION DIODE; FOUR-HETEROJUNCTION DIODE; MONORAIL DOUBLE-HETEROJUNCTION DIODE.*

hierarchically synchronized network A mutually synchronized network in which some clocks exert more control than others, the network operating frequency being a weighted mean of the natural frequencies of the population of clocks. *See also: OLIGARCHICALLY SYNCHRONIZED NETWORK.*

high-level control In data transmission, the conceptual level of control or processing logic existing in the hierarchical structure of a primary or secondary station that is above the link level and upon which the performance of link level functions are dependent or are controlled, e.g., device control, buffer allocation, or station management.

high-level data link control (HDLC) High-level control applied to a data link. *See also: DATA TRANSMISSION; LINK.*

high-loss fiber An optical fiber having a high energy loss, due to all causes, per unit length of fiber, usually measured in dB/km at a specified wavelength.
Note: High-loss is usually considered to be above 100 dB/km attenuation in amplitude of a propagating wave, caused primarily by scattering due to metal ions and by absorption due to water in the OH radical form.

highly-reflective coating A broad class of single or multilayer coatings that are applied to an optical surface for the purpose of increasing its reflectance over a specified range of wavelengths. Single films of aluminum or silver are common; but multilayers of at least two dielectrics are utilized when low absorption is imperative. Other parameters, such as angle of incidence, and intensity of radiation are also significant.

high-pass filter A filter that passes frequencies above a given frequency and attenuates all others.

hole Solid-state physics concept for a vacant electron site in the valence band. Such a vacant site is created when introducing an acceptor-type impurity in the semiconductor, or by injection from one type of semiconductor to another. When an electric field is applied, a "hole" behaves like a positively charged particle.

homochronous The relationship between two signals such that their corresponding significant instants are displaced by a constant interval of time. *See also: ANISOCHRONOUS; HETEROCHRONOUS; ISOCHRONOUS; MESOCHRONOUS; PLESIOCHRONOUS; SYNCHRONOUS SYSTEM.*

homogeneous cladding That part of the cladding wherein the refractive index is constant within a specified tolerance, as a function of radius. *See also: CLADDING; TOLERANCE FIELD.*

homogeneous multiplexing Multiplexing in which all the information bearer channels operate at the same data signaling rate. *See also: MULTIPLEX.*

homojunction A junction between semiconductors that may differ in their doping level conductivities but not in their atomic or alloy compositions. *See also: HETEROJUNCTION.*

hop The excursion of a radio wave from the earth to the ionosphere and back to the earth.
Note: The number of hops indicate the number of reflections from the ionosphere.

horizon angle The angle, in a vertical plane, between a horizontal line extending from the center of the antenna and a line extending from the same point to the radio horizon.

hour *See: LUMEN-HOUR.*

housing *See: LASER PROTECTIVE HOUSING.*

hunting 1. Pertaining to the operation of a selector or other similar device to find and establish connection with an idle circuit of a chosen group. 2. Pertaining to the failure of a device to achieve a state of equilibrium, usually by alternately overshooting or undershooting the point of equilibrium. *See also: FAILURE.*

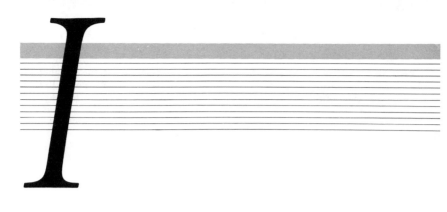

Iceland spar A transparent variety of the natural uniaxial crystal calcite that displays very strong double refraction, chemically being calcium carbonate crystallized in the hexagonal rhombohedral crystallographic system. *Synonym: CALSPAR.*

ideal blackbody *See: BLACKBODY.*

idle-channel noise Noise which is present in a communications channel when no signals are applied to it. The conditions and termination must be stated for the value to be significant.

IFS *Acronym for IONOSPHERIC FORWARD SCATTER.*

ILD *Acronym for INJECTION LASER DIODE.*

illuminance Luminous flux incident per unit area of a surface.

Note: Illuminance is expressed in lumens per square meter. *Synonym: ILLUMINATION.*

illumination *See: ILLUMINANCE.*

image In an optical system, a representation of an object produced by means of light rays.

Note: An image-forming optical element forms an image by collecting a bundle of light rays diverging from an object point and transforming it into a bundle of rays that converge toward, or diverge from, another point. If the rays converge to a point a real image of the object point is formed; if the rays diverge without intersecting each other they appear to proceed from a virtual image. *See also: DOUBLE IMAGE; ERECT IMAGE; GEOMETRIC IMAGE; GHOST IMAGE; REFLECTION IMAGE; REVERTED IMAGE; VIRTUAL IMAGE.*

image antenna A hypothetical, "mirror-image" antenna located as far below ground as the actual antenna is above ground.

image aspect The spatial orientation of an image, such as normal, canted, inverted, reverted, or rotated.

image brightness In an optical system, the apparent brightness of an image as seen through the optical system.

Note: This brightness depends on the brightness of the object, the transmission, magnification, distortion, and diameter of the exit pupil of the instrument.

image dissector In fiber optic systems, a bundle of fibers, with a tightly packed end on which an image may be focussed, in which the fibers may be separated into groups, each group transmitting part of the image, each part remaining coherent.

image intensifier A device, such as an electro-optic tube with a fiber-optic faceplate, capable of increasing the luminance of a low-intensity image or source.

image inverter In fiber optic systems, an image rotator that rotates the image 180 degrees.

image jump The apparent displacement of an object due to a prismatic condition in an optical system.

image plane The plane in which an image lies or is formed, perpendicular to the axis of a lens. A real image formed by a converging lens would be visible upon a screen placed in this plane.

image quality Those properties of a lens or optical system that affect the optical performance, such as resolving power, aberrations, image defects, and contrast rendition.

Note: Aberrations contribute to poor image quality. Errors of construction and defects in materials adversely affect image quality. Characteristic effects of aberrations on image quality make it possible to distinguish between their effects and those of accidental errors of workmanship, such as nonspherical surfaces, poor polish, scratches, pits, decentering, defects in cementing, and scattered light, all of which contribute to deterioration of the image. Defects in glass such as bubbles, stones, striae, crystalline bodies, cloudiness, strain, seeds, chicken-wire, and opaque minerals play a part in poor image quality.

image rotator   In a fiber optic system, a coherent bundle of fibers, the output end of which can be rotated with respect to the input end, thus twisting the bundle along its length and rotating the output image.

impedance   The total passive opposition offered to the flow of an alternating current. It consists of a combination of resistance, inductive reactance, and capacitive reactance. *See also: ITERATIVE IMPEDANCE, TERMINAL IMPEDANCE.*

impedance matching   The connection of an additional impedance to an existing one in order to improve performance or to accomplish a specific effect.

improvement threshold   *See: FM IMPROVEMENT THRESHOLD.*

impulse   A surge of electrical energy, usually of short duration, of a nonrepetitive nature. *See also: PULSE.*

impulse excitation   The production of an oscillatory current in a circuit by impressing a voltage for a relatively short period compared with the duration of the current produced. *Synonym: SHOCK EXCITATION.*

impulse noise   Noise consisting of random occurrences of energy spikes, having random amplitude and bandwidths, whose presence in a data channel can be a prime cause of errors. *See also: NOISE.*

impulse response   The function h(t) describing the response of an initially relaxed system to an impulse (Dirac-delta) function applied at time t = 0. The root-mean-square (rms) duration, σ_{rms}, of the impulse response is often used to characterize a component or system through a single parameter rather than a function.

$$\sigma_{rms} = [1/M_0 \int_{-\infty}^{\infty} (T-t)^2 h(t)dt]^{1/2}$$

$$\text{where } M_0 = \int_{-\infty}^{\infty} h(t)\, dt$$

$$T = 1/M_0 \int_{-\infty}^{\infty} th(t)\, dt.$$

Note: The impulse response may be obtained by deconvolving the input waveform from the output waveform, or as the inverse Fourier transform of the transfer function. *See also: ROOT-MEAN-SQUARE (RMS) PULSE DURATION; TRANSFER FUNCTION.*

impurity absorption   In lightwave transmission media, such as optical fibers and integrated optical circuits made of glass, silica, plastic, and other materials, the absorption of light energy from a traveling or standing wave by foreign elements in the medium, such as iron, copper, vanadium, chromium, hydroxide, and chloride ions.

Note: If the Fe, Cu, V, and Cr concentrations can be held below 8, 9, 18, and 28 parts per billion respectively, less than 20 dB/km losses can be obtained at band center. Power absorption occurs predominantly from foreign substances, such as transition metal ions like iron, cobalt, and chromium. Slab dielectric waveguides are included.

impurity level   An electron energy level of a material outside the normal energy levels of the material, caused by the presence of impurity atoms in the material.

Note: Such levels are capable of making an insulator semiconducting. The impurity atom may be a donor or an acceptor. If a donor, the impurity induces electronic conduction through the transfer of an electron to the conduction band. If an acceptor, the impurity can induce hole conduction through the acceptance of an electron from the valence band.

In   *Chemical symbol for INDIUM.*

in-band noise power ratio   For multichannel equipment, the ratio of the mean noise power measured in any channel, with all channels loaded with white noise, to the mean noise power measured in the same channel, with all channels but the measured channel loaded with white noise. *See also: NOISE.*

incandescence   The emission of light by thermal excitation that brings about energy level transitions that produce quantities of photons sufficient to render the source of radiation visible.

incidence   The act of falling upon or affecting, as a ray of light upon a surface.

Note: The ray is in the direction of propagation and perpendicular to the wavefront, which contains the electric and magnetic vectors of a transverse electromagnetic wave. *See also: GRAZING INCIDENCE; NORMAL INCIDENCE.*

incidence angle   In optics, the angle between the normal to a reflecting or refracting surface and the incident ray.

incident ray   A ray of light that falls upon, or strikes, the surface of any object such as a lens, mirror, prism, this printed page, the things we see, or the human eye.
Note: It is said to be incident to the surface.

inclusion   Denoting the presence of extraneous or foreign material.

incoherent   Characterized by a degree of coherence significantly less than 0.88. *See also: COHERENT; DEGREE OF COHERENCE.*

independent-sideband transmission   That method of double-sideband transmission in which the information carried by each sideband is different.
Note: The carrier may be suppressed. *See also: SIDEBAND TRANSMISSION.*

index   *See: ABSOLUTE REFRACTIVE INDEX.*

index dip   A decrease in the refractive index at the center of the core, caused by certain fabrication techniques. Sometimes called profile dip. *See also: REFRACTIVE INDEX PROFILE.*

index fiber   *See: GRADED-INDEX FIBER; STEP-INDEX FIBER.*

Index-guided diode laser   Diode laser for which lateral mode control is obtained by the fabrication of a real-index dielectric structure in the plane of the junction. Examples: BH, CDH, CSP lasers. Index-guided devices operate mostly in a single longitudinal mode and thus are sought for single-mode fiber communication systems. Index-guided lasers have stigmatic Gaussian-shaped beams.

index matching material   A material, often a liquid or cement, whose refractive index is nearly equal to the core index, used to reduce Fresnel reflections from a fiber end face. *See also: FRESNEL REFLECTION; MECHANICAL SPLICE; REFRACTIVE INDEX.*

index of refraction   *Synonym: REFRACTIVE INDEX (OF A MEDIUM).*

index profile   In an optical waveguide, the refractive index as a function of radius. *See also: GRADED INDEX PROFILE; PARABOLIC PROFILE; POWER-LAW INDEX PROFILE; PROFILE DISPERSION;* *PROFILE DISPERSION PARAMETER; PROFILE PARAMETER; STEP INDEX PROFILE; UNIFORM INDEX PROFILE.*

indirect bandgap semiconductors   Semiconductor compounds for which photons are emitted by electron-hole recombination only with the help of lattice vibrations (i.e., phonons). The corresponding optical transitions are called indirect transitions. Indirect-bandgap material is not suitable for lasing, but with certain dopants it can provide efficient LEDs. *Examples:* Si; Ge; GaP; $Al_xGa_{1-x}As$ for $0.4 < x < 1.0$.

indium   Metal used for bonding cw diode lasers to heatsinks. Indium is a soft metal of relatively low melting point ($T_M = 156°C$) and thus provides stress-free bonding.

indium gallium arsenide   A crystalline semiconductor alloy between indium phosphide and indium arsenide. The compound $In_{0.53}Ga_{0.47}As$ is lattice-matched to InP and thus is grown in layer form for 1.7 μm LED or laser emission or for long-wavelength APD or PIN photo-detectors. Symbols: InGaAs, (In,Ga)As, $In_xGa_{1-x}As$.

indium gallium arsenide phosphide   A crystalline semiconductor alloy between indium phosphide, gallium arsenide, gallium phosphide and indium arsenide. Written as $In_{1-x}Ga_xAs_yP_{1-y}$. When $y \cong 2.2x$ the compound can be lattice-matched to InP and thus provides the light emitting medium for lasers and LEDs emitting in the 1.06 to 1.7 μm range. Also used for long-wavelength APD and PIN detectors. *See: LONG-WAVELENGTH SOURCES, QUATERNARY LASER AND/OR LEDS.* Symbols: InGaAsP, (In,Ga) (As,P), $In_{1-x}Ga_xAs_yP_{1-y}$.

indium phosphide   A binary semiconductor compound used as substrate and as cladding layer, for carrier and light confinement, in long-wavelength LEDs and lasers emitting in the 1.06 to 1.7 μm range. *See: QUATERNARY LEDs AND LASERS. See: LONG-WAVELENGTH DEVICES.*

individual normal magnification   The apparent magnification produced by a magnifier, such as a lens or a mirror, when a person has myopia or hyperopia (hypermetropia).
Note: The individual normal magnification may be different from the absolute magnification.

induced optical conductor loss *See:* CONNECTOR-INDUCED OPTICAL CONDUCTOR LOSS.

information bearer channel A channel provided for data transmission which is capable of carrying all the necessary information to permit communications, including user's data, synchronizing sequences, control signals, etc. It may, therefore, operate at a greater signaling rate than that required solely for the user data. *See also:* ENVELOPE; INFORMATION.

information bit A bit which is generated by the data source and delivered to the data sink and which is not used by the data transmission system.

information channel The transmission media and equipment used for transmission in a given direction between two terminals.

infrared absorption edge Intrinsic absorption in silica-based fibers due to fundamental vibrational bands in SiO_2. Referred to as IR edge. The IR edge sets the long-wavelength limit for transmission in silica-based fibers. It typically manifests itself by a sharp increase in overall fiber loss at wavelengths above ≅1.7 μm.
Note: For examples see D. Botez and G. Herskowitz, *Proc. IEEE*, vol. 68, pp. 689–732, June 1980.

infrared band The band of electromagnetic wavelengths between the extreme of the visible part of the spectrum, about 0.75 microns, and the shortest microwaves, about 1000 microns.
Note: The IR region is sometimes subdivided into near infrared, 0.75-3 microns; middle infrared, 3-30 microns; and far infrared, 30-1000 microns.

infrared-emitting diode A semiconductor device in which radiative recombination of injected minority carriers produces infrared radiant flux when current flows as a result of applied forward voltage.

infrared (IR) The region of the electromagnetic spectrum between the long-wavelength extreme of the visible spectrum (about 0.7 μm) and the shortest microwaves (about 1 mm).

InGaAs *Chemical symbol for* INDIUM GALLIUM ARSENIDE.

InGaAsP *Chemical symbol for* INDIUM GALLIUM ARSENIDE PHOSPHIDE.

injection fiber *Synonym:* LAUNCHING FIBER.

injection laser *See:* SEMICONDUCTOR LASER.

injection laser diode (ILD) A semiconductor laser employing carrier injection across a forward-biased p–n junction as the means for pumping the active medium. *Synonyms:* DIODE LASER; SEMICONDUCTOR LASER. *See also:* ACTIVE LASER MEDIUM; CHIRPING; LASER; SUPERRADIANCE.

InP *Chemical symbol for Indium Phosphide.*

insertion gain The gain resulting from the insertion of a transducer in a transmission system, expressed as the ratio of the power delivered to that part of the system following the transducer to the power delivered to that same part before insertion. If more than one component is involved in the input or output, the particular component used must be specified. This ratio is usually expressed in decibels. If the resulting number is negative, an insertion loss is indicated.

insertion loss The total optical power loss caused by the insertion in the light path of an optical component such as a connector, splice, coupler, or modulator/switch.

insertion loss vs. frequency characteristic The amplitude transfer function characteristic of a system or component as a function of frequency. The amplitude response may be stated as actual gain, loss, amplification or attenuation, or as a ratio of any one of these quantities, at a particular frequency, with respect to that at a specified reference frequency. *Synonyms:* AMPLITUDE FREQUENCY RESPONSE; FREQUENCY RESPONSE. *See also:* FREQUENCY; TRANSFER FUNCTION.

inside vapor-phase oxidation process (IVPO) A CVPO process, for production of optical fibers, in which dopants, such as silicon tetrachloride, are burned with dry oxygen and a fuel gas to form an oxide (soot) stream which is deposited on the inside of a rotating glass tube, then sintered to produce a doped layer of higher refractive index glass on the inside. The tube then being drawn into a solid fiber. *See also:* OUTSIDE VAPOR PHASE OXIDATION PROCESS; CHEMICAL VAPOR PHASE OXIDATION PROCESS; MODIFIED INSIDE VAPOR-PHASE OXIDATION PROCESS.

integrated-optical circuit filter-coupler-switch-modulator Two or more optical waveguides fabricated on a

minute piece of material, such as lithium niobate, whose light-propagating characteristics and energy interaction can be varied, such as by applying a voltage across a common section of the waveguides, so as to perform the four major functions found in a radio receiver, namely filtering, coupling, switching, and modulating.
Note: Special electrodes control the performance of the various functions.

integrated optical circuit (IOC) An optical circuit, either monolithic or hybrid, composed of active and passive components, used for coupling between optoelectronic devices and providing signal processing functions.

integrated optics The interconnection of miniature optical components via optical waveguides on transparent dielectric substrates, using optical sources, modulators, detectors, filters, couplers, and other elements incorporated into circuits analagous to integrated electronic circuits for the execution of various communication, switching, and logic functions.

intensifier *See: IMAGE INTENSIFIER.*

intensity The square of the electric field amplitude of a light wave. Intensity is proportional to irradiance and may be used in place of the term "irradiance" when only relative values are important. *See also: IRRADIANCE; RADIANT INTENSITY; RADIOMETRY.*

intensity modulation *See: ANALOG-INTENSITY MODULATION.*

interaction crosstalk Crosstalk caused by coupling between carrier and non-carrier circuits; the crosstalk may in turn be coupled to another carrier circuit. *See also: COUPLING.*

interference In optics, the interaction of two or more beams of coherent or partially coherent light. *See also: COHERENT; DEGREE OF COHERENCE; DIFFRACTION.*

interference filter An optical filter consisting of one or more thin layers of dielectric or metallic material. *See also: DICHROIC FILTER; INTERFERENCE; OPTICAL FILTER.*

interferometer An instrument that employs the interference of light waves for purposes of measurement. *See also: INTERFERENCE.*

intermediate field *Synonym: INTERMEDIATE FIELD REGION.*

intermediate field region The transition region, lying between the near field region and the far field region, in which the electric field strength of an electromagnetic wave developed by a transmitting antenna is dependent upon the inverse distance, inverse square of the distance, and the inverse cube of the distance from the antenna.
Note: The intermediate field region is considered as any distance between 0.1 and 1.0 of the wavelength, for an antenna equivalent length that is small compared to this distance. *Synonyms: INTERMEDIATE FIELD; INTERMEDIATE ZONE. See Also: FAR FIELD REGION; NEAR FIELD REGION.*

intermediate zone *Synonym: INTERMEDIATE FIELD REGION.*

intermodal distortion *Synonym: MULTIMODE DISTORTION.*

intermodulation The production, in a nonlinear transducer element, of frequencies corresponding to the sums and differences of the fundamentals and harmonics of two or more frequencies which are transmitted through the transducer.

intermodulation distortion Nonlinear distortion characterized by the appearance of frequencies in the output, equal to the sums and differences of integral multiples of the component frequencies present in the input.
Note: Harmonic components also present in the output are usually not included as part of the intermodulation distortion. When harmonics are included, a statement to that effect should be made.

intermodulation noise In a transmission path or device, that noise which is contingent upon modulation and demodulation and results from any nonlinear characteristics in the path or device. *See also: EQUIPMENT INTERMODULATION NOISE.*

internal absorptance The ratio of light flux absorbed between the entrance and emergent surfaces of a medium, to the flux that has penetrated the entrance surface.
Note: The effects of interreflections between the two surfaces are not included. Internal absorptance is numerically equal to unity minus the internal transmittance.

internal optical density The logarithm to the base ten of the reciprocal of the internal transmittance. *Synonym: TRANS-MISSION FACTOR.*

internal photoeffect The changes in characteristics of a material that occur, such as conductivity, emissivity, or electric potential developed, when incident photons are absorbed by the material and excite the electrons in the various energy bands. For example, electrons may move from a valence band to a conduction band for both intrinsic material and impurities; or to or from the valence bands of the intrinsic material and impurities; i.e. dopants and other impurities; thus both intrinsic and extrinsic photoeffects may be involved in the internal photoeffect. *See also: EXTRINSIC INTERNAL PHOTOEFFECT; INTRINSIC INTERNAL PHOTOEFFECT.*

internal photoeffect detector A photodetector in which incident photons raise electrons from a lower to a higher energy state. Resulting in an altered state of the electrons, holes, or electron-hole pairs generated by the transition, which is then detected.
Note: Most semiconductors make use of the internal photoeffect for signal detection at the end of an optical fiber. *See also: PHOTOEFFECT DETECTOR.*

internal quantum efficiency The fraction of excited carriers in a semiconductor device that recombine to provide light. The internal quantum efficiency is a measure of the material quality for devices operated in the spontaneous emission mode (i.e., LEDs). In high-quality direct-bandgap semiconductors, such as GaAs, the internal quantum efficiency is close to unity.
Note: The internal quantum efficiency in stimulated emission (i.e., lasing) is always close to unity, while the internal quantum efficiency in spontaneous emission can be well below unity, such as in long-wavelength devices. Symbol: η_i.

internal reflection *See: TOTAL INTERNAL REFLECTION.*

internal reflection angle *See: CRITICAL ANGLE.*

internal transmittance The ratio of the flux transmitted to the second surface of a medium to the corresponding flux that has just passed through the first surface, i.e. the transmittance from the first surface to the second surface.
Note: Internal transmittance does not include the effects due to interreflection between the two surfaces.

interrupted isochronous Pertaining to isochronous burst transmission.

interrupted isochronous transmission *Synonym: ISOCHRONOUS BURST TRANSMISSION.*

intersymbol interference Extraneous energy from the signal in one or more keying intervals that tends to interfere with the reception of the signal in another keying interval, or the disturbance that results therefrom.

intramodal distortion That distortion resulting from dispersion within individual propagating modes. It is the only distortion occurring in single mode waveguides. *See also: DISPERSION; DISTORTION.*

intrinsic absorption In lightwave transmission media, such as optical fibers and integrated optical circuits made of glass, silica, plastic, and other materials, the absorption of light energy from a traveling or standing wave by the medium itself, causing attenuation as a function of distance, material properties, mode, frequency, and other factors.
Note: Intrinsic absorption is primarily due to charge transfer bands in the ultraviolet region and vibration or multiphonon bands in the near infrared, particularly if they extend into the region of wavelengths used in fiber communications, namely 700–1100 nanometers.

intrinsic internal photoeffect An internal photoeffect involving the basic material rather than any dopants or other impurities. *See also: EXTRINSIC INTERNAL PHOTOEFFECT.*

intrinsic joint loss That loss, intrinsic to the fiber, caused by fiber parameter (e.g., core dimensions, profile parameter) mismatches when two nonidentical fibers are joined. *See also: ANGULAR MISALIGNMENT LOSS; EXTRINSIC JOINT LOSS; GAP LOSS; LATERAL OFFSET LOSS.*

intrinsic junction loss The total loss resulting from joining two identical optical waveguides.
Note: Factors influencing this loss include spacing loss, alignment of the waveguides, Fresnel reflection loss, end finish, etc. *See also: ANGULAR MISALIGNMENT LOSS; GAP LOSS; EXTRINSIC JUNCTION LOSS; LATERAL OFFSET LOSS.*

intrinsic-negative photodiode coupler
See: POSITIVE-INTRINSIC-NEGATIVE PHOTODIODE COUPLER.

intrinsic noise   In a transmission path or device, that noise which is inherent to the path or device and is not contingent upon modulation.

inversion   *See: POPULATION INVERSION.*

inverted position   In frequency division multiplexing, a position of a translated channel in which an increasing signal frequency in the untranslated channel causes a decreasing signal frequency in the translated channel.

inverted-rib-waveguide (IRW) diode laser   Mode-stablized LOC-type, long-wavelength diode laser structure grown by one-step liquid-phase epitaxy over a channel etched into a non-absorbing InP substrate. The guide layer has a local increase in thickness, where it fills in the channel. The active layer is of constant thickness. Typical reliable output power levels: 3–5 mW/facet CW.
Note: For example see S. Turley, *et al.*, *Electron. Lett.*, vol. 17, pp. 868–870, 1981.

IOC   *Acronym for INTEGRATED OPTICAL CIRCUIT.*

ion exchange technique   A method of fabricating a graded index optical waveguide by an ion exchange process. *See also: CHEMICAL VAPOR DEPOSITION TECHNIQUE; DOUBLE CRUCIBLE METHOD; GRADED INDEX PROFILE.*

ion laser   A gas laser involving ionization of certain gases, such as argon, krypton, and xenon.

ionosphere   The region of the atmosphere, extending from roughly 50 km to 400 km altitude, in which there is appreciable ionization. The presence of charged particles in this region profoundly affects the propagation of electromagnetic radiations of long wavelength (radio and radar waves).
Note: 1. The ionosphere includes the following layers or regions:
D: Between about 50 km to 90 km altitude; responsible for most of the attenuation of radio waves in the range 1 MHz to 100 MHz. Exists only during daylight hours.
E: At about 110 km altitude; formerly called Heaviside layer (or Kennelly-Heaviside).
F: The region between about 175 km to 400 km altitude: F_1, at about 175 km to

250 km, exists only in daylight hours. F_2, at about 250 km to 400 km is the principal reflecting region for long-distance HF communication.
Note: 2. The term may also be used with reference to other planets. *See also: IONOSPHERIC DISTURBANCE; IONOSPHERIC SCATTER; SKYWAVE; SPORADIC E PROPAGATION; SUDDEN IONOSPHERIC DISTURBANCE.*

ionospheric disturbance   A sudden increase in the ionization of the D-region of the ionosphere, caused by solar flares, which results in greatly increased radio wave absorption. *See also: IONOSPHERE; SUDDEN IONOSPHERIC DISTURBANCE.*

ionospheric forward scatter (IFS)   *Synonym: IONOSPHERIC SCATTER.*

ionospheric scatter   The propagation of radio waves by scattering due to irregularities in the ionosphere. *Synonyms: IONOSPHERIC FORWARD SCATTER (IFS); FORWARD PROPAGATION IONOSPHERIC SCATTER (FPIS). See also: IONOSPHERE; PROPAGATION.*

IR   *Acronym for INFRARED.*

IR edge   *See: INFRARED ABSORPTION EDGE.*

irradiance   Radiant power incident per unit area upon a surface, expressed in watts per square meter. "Power density" is colloquially used as a synonym. *See also: RADIOMETRY.*

irradiation   The product of irradiance and time, therefore radiant energy received per unit area.

IRW laser   *See: INVERTED-RIB-WAVEGUIDE DIODE LASER.*

isochronous   1. That characteristic of a periodic signal in which the time interval separating any two corresponding transitions is equal to the unit interval or to a multiple of the unit interval. 2. Pertaining to data transmission in which corresponding significant instants of two or more sequential signals have a constant phase relationship. *See also: ANISOCHRONOUS; ASYNCHRONOUS TRANSMISSION; HETEROCHRONOUS; HOMOCHRONOUS; MESOCHRONOUS; PLESIOCHRONOUS.*

isochronous burst transmission   A transmission process which may be used where the information bearer channel rate is higher than the input data signaling rate. The binary digits being transferred are signaled at the digit rate of the information bearer channel rate, and the transfer is interrupted at intervals in order to produce the required mean data signaling rate. The interruption is always for an intergral number of digit periods. *Note:* The isochronous burst condition has particular application where envelopes are being transmitted and received by the data circuit-terminating equipment, but only the bytes contained within the envelopes are being transferred between data circuit-terminating equipment and the data terminal equipment. *Synonyms: BURST ISOCHRONOUS (Deprecated); INTERRUPTED ISOCHRONOUS TRANSMISSION. See also: ENVELOPE; INFORMATION TRANSFER.*

isochronous distortion   The difference between the measured modulation rate and the theoretical modulation rate in a digital system. *See also: DEGREE OF ISOCHRONOUS DISTORTION.*

isochronous modulation   Modulation (or demodulation) in which the time interval separating any two significant instants is theoretically equal to the unit interval or to a multiple of the unit interval. *See also: MODULATION; PLESIOCHRONOUS.*

isolator   A device intended to prevent return reflections along a transmission path.

Note. The Faraday isolator uses the magneto-optic effect.

isotropic   Pertaining to a material whose electrical or optical properties are independent of direction of propagation and of polarization of a traveling wave. *See also: ANISOTROPIC; BIREFRINGENT MEDIUM.*

isotropic antenna   A hypothetical antenna that radiates or receives equally in all directions.

Note: Isotropic antennas do not exist physically but represent convenient reference antennas for expressing directional properties of actual antennas.

isotropic material   A substance that exhibits the same property when tested along an axis in any direction, for example, a dielectric material with the same permittivity or a glass with the same index of refraction in all directions.

iterative impedance   At a pair of terminals of a four-terminal network, the impedance that will terminate the other pair of terminals in such a way that the impedance measured at the first pair is equal to this terminating impedance. The iterative impedance of a uniform line is the same as its characteristic impedance.

IVPO   *Acronym for INSIDE VAPOR-PHASE OXIDATION PROCESS.*

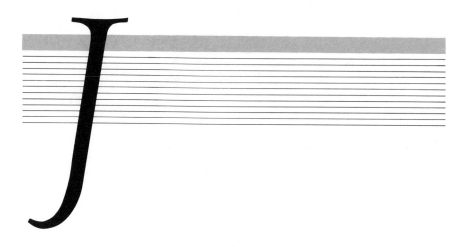

jacket *See: BUNDLE JACKET; CABLE JACKET; OPTICAL FIBER JACKET.*

jitter Abrupt and spurious variations in a signal, such as in interval duration, amplitude of successive cycles, or in the frequency or phase of successive pulses. *Note:* When used qualitatively, the term must be identified as being time-, amplitude-, frequency-, or phase-related and the form must be specified, e.g., pulse-delay-time jitter, pulse-duration jitter. When used quantitatively, a measure

Johnson noise *Synonym: THERMAL NOISE.*

jump *See: IMAGE JUMP.*

junction *See: DOUBLE HETEROJUNCTION; HETEROJUNCTION; SINGLE HETEROJUNCTION; OPTICAL FIBER JUNCTION.*

of the time- or amplitude-related variation must be included, e.g., average rms, peak-to-peak. *See also: PHASE JITTER; PHASE PERTURBATION; TIME JITTER.*

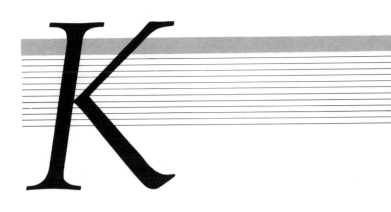

k *Symbol: BOLTZMANN'S CONSTANT.*

Kelvin temperature scale In the International System of Units (SI), the kelvin (K) is defined as the fraction 1/273.16 of the thermodynamic temperature of the triple point of water. The temperature 0 K is called "absolute zero" (equivalent to -273.16°C or -459.69°F). The degree Celsius is now defined as an interval of one kelvin, rather than "one degree Kelvin." *Formerly called ABSOLUTE TEMPERATURE SCALE.*

Kerr cell A substance, usually a liquid, whose refractive index change is proportional to the square of the applied electric field, the substance being configured so as to be part of another system, such as an optical path, the cell thus providing a means of modulating the light in the optical path.

kink Strong non-linearity in the light output vs. current input of a diode laser. The "kink" usually indicates a shift of the lasing spot and/or the laser beam, and is frequently associated with excess noise.

Note: Mode-stabilized diode lasers do not have kinks in their light vs. current characteristics.

Note: Lasers that do not have kinks in their light vs. current characteristics are not necessarily operating in a single spatial mode.

knife edge effect The transmission of radio signals into the line-of-sight shadow region caused by the diffraction over an obstacle, e.g., a sharply defined mountain top.

knife-edge test *See: FOUCAULT KNIFE-EDGE TEST.*

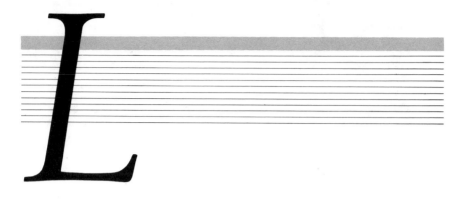

lambert A unit of luminance equal to $\frac{10^4}{\pi}$ candles per square meter.

Note: The SI unit of luminance is the lumen per square meter wherein 4π lumens of light flux emanate from one candela.

Lambertian Pertaining to a radiance distribution that is uniform in all directions. *See also: UNIFORM LAMBERTIAN.*

Lambertian radiator *See: LAMBERT'S COSINE LAW.*

Lambertian reflector *See: LAMBERT'S COSINE LAW.*

Lambertian source *See: LAMBERT'S COSINE LAW.*

Lambert's emission law *See: COSINE EMISSION LAW.*

Lambert's law In the transmission of electromagnetic radiation through a scattering or absorptive medium, the internal transmittance of a given thickness D_2 is related to the known transmittance, T_1 of a known thickness, D_1, by the relationship:

$$T_2 = T_1^{(D_2/D_1)}$$

See also: BEER'S LAW; BOUGER'S LAW.

Lambert's cosine law The statement that the radiance of certain idealized surfaces, known as Lambertian radiators, Lambertian sources, or Lambertian reflectors, is independent of the angle from which the surface is viewed.

Note: The radiant intensity of such a surface is maximum normal to the surface and decreases in proportion to the cosine of the angle from the normal. *Synonym: COSINE EMISSION LAW.*

large optical-cavity (LOC) diode A laser diode for which the dielectric waveguide contains both an active layer and a guide layer. The optical mode mostly propagates in the guide layer, while obtaining gain from the active layer. The guide layer is relatively thick (1-

to 2-μm) and thus allows a large spot size in the transverse junction.

laser A device that produces optical radiation using a population inversion to provide Light Amplification by Stimulated Emission of Radiation and (generally) an optical resonant cavity to provide positive feedback. Laser radiation may be highly coherent temporally, or spatially, or both. *See also: ACTIVE LASER MEDIUM; INJECTION LASER DIODE; OPTICAL CAVITY.*

laser basic mode The primary or lowest order fundamental transverse propagation mode for the emitted light wave of a laser; the emitted energy normally having Gaussian (bell-shaped) distribution in space and all the energy is in a single beam, with no side lobes.

laser diode coupler A coupling device that enables the coupling of light energy from a Laser Diode (LD) source to an optical fiber or cable at the transmitting end of an optical fiber data link.

Note: The coupler may be an optical fiber pigtail epoxied to the LD. *Synonym: LD COUPLER.*

laser diode (LD) *Synonym: INJECTION LASER DIODE.*

laser fiber optic transmission system A system consisting of one or more laser transmitters and associated fiber optic cables.

Note: During normal operation, the laser radiation is limited to the cable. Thus, laser systems that employ fiber optic transmission shall have cable service connection that requires a tool to disconnect if such cable connections form part of the protective housing. Consideration should also be given to incorporating mechanical beam attenuators at connectors. Safety aspects peculiar to fiber optics are an important consideration.

laser head A module containing the active laser medium, resonant cavity, and other components within one enclosure, not necessarily including a power supply.

laser linewidth In the operation of a laser, the frequency range over which most of the laser beam's energy is distributed.

laser medium *Synonym: ACTIVE LASER MEDIUM.*

laser protective housing A protective housing for a laser to prevent human exposure to laser radiation in excess of an allowable established, or statutory emission limit for the appropriate class. *Note:* Parts of the housing that can be removed or displayed and not interlocked may be secured in such a way that removal or displacement of the parts requires the use of special tools.

laser pulse length The time duration of the burst of electromagnetic energy emitted by a pulsed laser. *Note:* It is usually measured at the half-power points, i.e., on a plot of pulse power developed versus time, the time interval between the points that are at 0.5 of the peak of the power curve. *Synonym: LASER PULSE WIDTH.*

laser pulse width *See: LASER PULSE LENGTH.*

laser service connection An access point in a laser-to-fiber-optic transmission system that is designed for service and that, for safety, should require a tool to disconnect.

lasing A phenomenon that implies the generation of self-sustained oscillation of coherent light generated from an active medium (i.e., semiconductor, gas, or liquid) mainly by stimulated emission. The oscillation is self-sustained, since positive feedback provided by mirrors ensures that the round-trip optical gain is equal to the cavity in internal and external losses. *Note:* This action takes place in a laser.

lasing cavity The ensemble of light confining parts, such as dielectric waveguides and mirrors, that defines the resonator for a laser oscillator. In a diode laser the lasing cavity is formed by the dielectric waveguide confining the light laterally and transversely with respect to the active region, and the mirror facets confining the light longitudinally.

lasing cavity modes The discrete frequencies characterizing the lasing cavity (resonator) at which the laser can oscillate. Depending on the design and the injected carrier dynamics a laser can oscillate either in a single mode or in many modes. *See: SINGLE-MODE LASER; MULTI-MODE LASER.*

lasing medium *See: ACTIVE LASER MEDIUM.*

lasing spot size The widths of the lasing mode to $1/e^2$ points in intensity in the plane of the junction, and in the plane perpendicular to the junction and parallel to the emitting facet. The range of widths in the plane of the junction is 1 to 7 μm, depending on the strength of the lateral mode confinement. In the plane perpendicular to the junction the mode width is 0.3 to 0.5 μm for DH-type lasers, and 0.5 to 2 μm for LOC-type lasers. *Note:* For AlGaAs devices the larger the spot size the higher the power capability, since reliability is a function of optical flux density at the emitting facet. That is not a factor in NAM-type devices.

lasing threshold The excitation level in a light-amplifying medium at which the cavity round-trip optical gain is equal to the cavity losses. At lasing threshold the light-amplifying device becomes a light oscillator. *See also: LASER; SPONTANEOUS EMISSION; STIMULATED EMISSION.*

lateral displacement loss In an in-line (butt) splice of an optical fiber, the loss of signal power caused by a side-wise displacement of the optical axes of the two fiber ends that are joined.

lateral magnification The ratio of the linear size of an image to that of the object, as when an enlarging lens is used.

lateral mode control The fabrication procedures via which stable lateral wave confinement (i.e., in the plane of the junction) is achieved. A diode laser with lateral mode control is a mode-stabilized device.

lateral offset loss A power loss caused by transverse or lateral deviation from optimum alignment of source to optical waveguide, waveguide to waveguide, or waveguide to detector. *Synonym: TRANSVERSE OFFSET LOSS.*

lateral wave confinement The mechanism and/or dielectric waveguiding structure that provides light confinement in the plane of the junction of a diode laser, in a direction parallel to the mirror facets.

launch angle The angle between the light input propagation vector and the optical axis of an optical fiber or fiber bundle. *See also: LAUNCH NUMERICAL APERTURE.*

launching fiber A fiber used in conjunction with a source to excite the modes of another fiber in a particular fashion.
Note: Launching fibers are most often used in test systems to improve the precision of measurements. *Synonym: INJECTION FIBER. See also: MODE; PIGTAIL.*

launch numerical aperture (LNA) The numerical aperture of an optical system used to couple (launch) power into an optical waveguide.
Note 1. LNA may differ from the stated NA of a final focusing element if, for example, that element is underfilled or the focus is other than that for which the element is specified.
Note 2. LNA is one of the parameters that determine the initial distribution of power among the modes of an optical waveguide. *See also: ACCEPTANCE ANGLE; LAUNCH ANGLE.*

law *See: BEER'S LAW; BOUGER'S LAW; BREWSTER'S LAW; COSINE-EMISSION LAW; LAMBERT'S LAW; PLANCK'S LAW; REFLECTION LAW; RICHARDSON'S LAW; SNELL'S LAW.*

LD *Acronym for LASER DIODE.*

LD coupler *See: LASER DIODE COUPLER.*

leakage current *See: PHOTODIODE LEAKAGE CURRENT.*

leakage loss *See: LIGHT-LEAKAGE LOSS.*

leaky mode In an optical waveguide, a mode whose field decays monotonically for a finite distance in the transverse direction but which becomes oscillatory everywhere beyond that finite distance. Specifically, a mode for which

$$[n^2(a)k^2 - (l/a)^2]^{1/2} \leq \beta \leq n(a)k$$

where β is the imaginary part (phase term) of the axial propagation constant, l is the azimuthal index of the mode, $n(a)$ is the refractive index at $r=a$, the core radius, and k is the free-space wavenumber, $2\pi/\lambda$, and λ is the wavelength. Leaky modes correspond to leaky rays in the terminology of geometric optics.
Note: Leaky modes experience attenuation, even if the waveguide is perfect in every respect. *Synonym: TUNNELING MODE. See also: BOUND MODE; CLADDING MODE; LEAKY RAY; MODE; UNBOUND MODE.*

leaky ray In an optical waveguide, a ray for which geometric optics would predict total internal reflection at the core boundary, but which suffers loss by virtue of the curved core boundary. Specifically, a ray at radial position r having direction such that

$$n^2(r) - n^2(a) \leq \sin^2\theta(r)$$

and

$$\sin^2\theta(r) \leq [n^2(r) - n^2(a)] / [1 - (r/a)^2\cos^2\phi(r)]$$

where $\theta(r)$ is the angle the ray makes with the waveguide axis, $n(r)$ is the refractive index, a is the core radius, and $\phi(r)$ is the azimuthal angle of the projection of the ray on the transverse plane. Leaky rays correspond to leaky (or tunneling) modes in the terminology of mode descriptors. *Synonym: TUNNELING RAY. See also: BOUND MODE; CLADDING RAY; GUIDED RAY; LEAKY MODE.*

leaky waveguide *See: ANTIGUIDE.*

LEA laser *See: LONGITUDINALLY EXCITED ATMOSPHERE LASER.*

least-time principle *See: FERMAT PRINCIPLE.*

LED *Acronym for LIGHT EMITTING DIODE.*

LED coupler *See: LIGHT-EMITTING DIODE COUPLER.*

length *See: BACK FOCAL LENGTH; EQUIVALENT FOCAL LENGTH; FRONT FOCAL LENGTH; FOCAL LENGTH; LASER PULSE LENGTH; OPTICAL PATH LENGTH; PULSE LENGTH.*

lens 1. An optical component made of one or more pieces of a material transparent to the radiation passing through, having curved surfaces, that is capable of forming an image, either real or virtual, of the object source of the radiation, at least one of the curved surfaces being convex or concave, normally spherical but sometimes aspheric. *See also: COLLECTIVE LENS; COMPOUND LENS; CONVERGING LENS; DIVERGING LENS.* 2. A transparent optical element, usually made from optical glass, having two opposite polished major surfaces of which at least one is convex or concave in shape and usually spherical.
Note: The polished major surfaces are shaped so that they serve to change the amount of convergence or divergence of the transmitted rays. *See also: ACHRO-*

MATIC LENS; APLANATIC LENS; BI-TORIC LENS; CARTESIAN LENS; CONCENTRIC LENS; CONDENSING; CORRECTED LENS; CYLINDRICAL LENS; DIVERGENT MENISCUS LENS; FIELD LENS; FINISHED LENS; PLANO LENS; PLANOCONCAVE LENS; PLANOCONVEX LENS; TAPERED LENS; TELEPHOTO LENS; THICK LENS; THIN LENS; ZOOM LENS.

lens coupling In optical waveguides, such as optical fibers and integrated optical circuits, the transfer of electromagnetic energy from source to guide, or from guide to guide, by means of a lens placed between the source and sink.
Note: Coupling loss can be reduced to packing fraction loss, axial misalignment loss and axial displacement loss when a lens is used. *See also: DIRECT COUPLING.*

lens measure A mechanical device for measuring surface curvature in terms of dioptric power.

lens speed That property of a lens that affects the illuminance of the image it produces.
Note: Lens speed may be specified in terms of the aperture ratio, numerical aperture, T-STOP, or F-NUMBER.

lens system Two or more lenses arranged to work in conjunction with one another.

lens watch A dial depth gauge graduated in diopters.

level *See: HIGH LEVEL DATA LINK CONTROL; ENERGY LEVEL; IMPURITY LEVEL.*

lever *See: OPTICAL LEVER.*

L-I curve Characteristic of diode laser or LED displaying the output light power vs. the input drive current. *Synonyms: LIGHT-CURRENT CHARACTERISTICS; P-I CURVE; POWER-CURRENT CHARACTERISTIC.*

light 1. In a strict sense, the region of the electromagnetic spectrum that can be perceived by human vision, designated the visible spectrum and nominally covering the wavelength range of 0.4μm to 0.7μm. 2. In the laser and optical communication fields, custom and practice have extended usage of the term to include the much broader portion of the electromagnetic spectrum that can be handled by the basic optical techniques used for the visible spectrum. This region has not been clearly defined but, as employed by most workers in the field,

may be considered to extend from the near-ultraviolet region of approximately 0.3μm, through the visible region, and into the mid-infrared region to 30μm. *See also: INFRARED (IR); OPTICAL SPECTRUM; ULTRAVIOLET (UV).*

light absorption The conversion of light into other forms of energy upon traversing a medium, thus weakening the transmitted light beam. Energy reflectance R, transmittance T, absorption A, and scattering S, obey the law of the conservation of energy:

$$R + T + A + S = 1$$

light adaptation The ability of the human eye to adjust itself to a change in the intensity of light.

light analyzer For incident light, a polarizing element that can be rotated about its axis to control the amount of transmission of incident plane polarized light, or to determine the plane of polarization of the incident light.

light antenna A system of reflecting and refracting optical components arranged to guide or direct a beam of light.

light conduit *See: NONCOHERENT BUNDLE.*

light current *See: PHOTOCURRENT.*

light-emitting diode coupler A coupling device that enables the coupling of light energy from a light-emitting diode (LED) source to an optical fiber or cable at the transmitting end of an optical fiber data link.
Note: The coupler may be an optical fiber pigtail epoxied to the LED. *Synonym: LED COUPLER.*

light emitting diode (LED) A pn junction semiconductor device that emits incoherent optical radiation when biased in the forward direction. *See also: INCOHERENT.*

lightguide *Synonym: OPTICAL WAVEGUIDE.*

light-leakage loss Light energy loss in a light transmission system, such as in a light conduit, optical fiber cable, connector, or optical integrated circuit, due to any means of escape, such as imperfections at core-cladding boundaries, breaks in jackets, and less-than-critical-radius bending.

light pencil In optics, a narrow bundle of light rays, diverging from a point source or converging toward an image point.

light quantity The product of luminous flux and time.

light ray The path of a point on a wavefront. The direction of a light ray is generally normal to the wavefront. *See also: GEOMETRIC OPTICS.*

lightwave communications That aspect of communications and telecommunications devoted to the development and use of equipment that uses electromagnetic waves in or near the visible region of the spectrum for communication purposes.
Note: Lightwave communication equipment includes sources, modulators, transmission media, detectors, converters, integrated optic circuits, and related devices, used for generating and processing light waves. The term optical communications is oriented toward the notion of optical equipment whereas the term lightwave communications is oriented toward the signal being processed. *Synonym:* OPTICAL COMMUNICA-TIONS. *See also: LIGHT.*

limit *See: ACCOMMODATION LIMIT.*

limited operation *See: DETECTOR-NOISE-LIMITED OPERATION; DISPER-SION-LIMITED OPERATION; QUAN-TUM-LIMITED OPERATION; THERMAL-NOISE-LIMITED OPERA-TION.*

limiter A device in which the power or some other characteristic of the output signal is automatically prevented from exceeding a specified value. *See also: CLIPPER; LIMITING; PEAK LIMITING.*

limiting A process by which some characteristic at the output of a device is prevented from exceeding a predetermined value.
Note: 1. Hard limiting is a limiting action with negligible variation in output in the range where the output is limited (controlled) when subjected to a fairly wide variation of signal input.
Note: 2. Soft limiting is a limiting action with appreciable variation in output in the range where the output is limited (controlled) when subjected to a fairly wide variation of signal input. *See also: LIMITER; PEAK LIMITING.*

limiting resolution angle The angle subtended by two points or lines that are just far enough apart to permit them to be distinguished as separate.
Note: The ability of an optical device to resolve two points or lines is called resolving power and quantitatively is inversely proportional to the limiting angle of resolution.

linear combiner A diversity combiner which adds two or more receiver outputs. *See also: DIVERSITY COMBINER.*

linearity A constant relationship, over a designated range, between input and output characteristics of a device. *See also: FIDELITY; NONLINEAR DISTORTION.*

linearly polarized (LP) mode A mode for which the field components in the direction of propagation are small compared to components perpendicular to that direction.
Note: The LP description is an approximation which is valid for weakly guiding waveguides, including typical telecommunication grade fibers. *See also: MODE; WEAKLY GUIDING FIBER.*

linear optical element A device for which the radiant output power is linearly proportional to the radiant input power and no new optical wavelengths or modulation frequencies are generated. A linear element can be described in terms of a transfer function or an impulse response function.

linear predictive coding (LPC) A narrowband analog-to-digital conversion technique employing a one-level or multilevel sampling system in which the value of the signal at each sample time is predicted to be a particular linear function of the past values of the quantized signal.
Note: LPC is related to adaptive predictive coding (APC) in that both use adaptive predictors. However, LPC uses more prediction coefficients to permit use of a lower information bit rate (about 2.4 to 4.8 kbps) than APC, and thus requires a more complex voice processor. *See also: ADAPTIVE PREDICTIVE CODING; CODE.*

LiNbO$_3$ *Chemical symbol for* LITHIUM NIOBATE.

line-of-sight (LOS) propagation Radio propagation in the atmosphere which is similar to light transmission in that intensity decreases mainly due to energy spreading according to the inverse-distance law with relatively minor effects due to the composition and structure of the atmosphere. Line-of-sight propagation is considered to be unavailable when any ray from the transmitting antenna, refracted by the atmosphere, will encounter the earth or any other opaque object that prevents the ray from proceeding directly to the receiving antenna.

line source 1. In the spectral sense, an optical source that emits one or more spectrally narrow lines as opposed to a continuous spectrum. *See also: MONO-CHROMATIC.* 2. In the geometric sense, an optical source whose active (emitting) area forms a spatially narrow line.

line spectrum An emission or absorption spectrum consisting of one or more narrow spectral lines, as opposed to a continuous spectrum. *See also: MONO-CHROMATIC; SPECTRAL LINE; SPEC-TRAL WIDTH.*

linewidth *See: SPECTRAL WIDTH.*

link 1. The communication facilities existing between adjacent nodes. 2. A portion of a circuit designed to be connected in tandem with other portions. 3. A radio path between two points, called a radio link, which may be unidirectional, half duplex, or full duplex.

Note: It is generally accepted that the signals at each end of a link are in the same form.

4. In computer programming, the part of a computer program, in some cases a single instruction or address, that passes control and parameters between separate portions of the computer program.

Note: The term link should be defined or qualified when used. *See also: CONTEN-TION; DATA LINK; OPTICAL DATA LINK; DOWNLINK; FUNCTIONAL SIG-NALING LINK; HIGH-LEVEL DATA LINK CONTROL; MULTIPOINT LINK; NONCENTRALIZED OPERATION; POINT-TO-POINT LINK; PRIMARY STATION; PROTOCOL; SYNCHRO-NOUS DATA-LINK CONTROL; TACTI-CAL DIGITAL INFORMATION LINK; UPLINK.*

link level In data transmission, the conceptual level of control or data processing logic existing in the hierarchical structure of a primary or secondary station that is responsible for maintaining control of the data link.

Note: The link level functions provide an interface between the station high level logic and the data link; these functions include transmit bit injection and receive bit extraction, address/control field interpretation, command response generation, transmission and interpretation, and frame check sequence computation and interpretation. *See also: DATA TRANSMISSION.*

link protocol A set of rules for data communication over a data link specified in terms of a transmission code, a transmission mode, and control and recovery procedures.

liquid-core fiber An optical fiber consisting of optical glass, quartz or silica tubing filled with a higher refractive index liquid, such as tetrachloroethylene that has attenuation troughs less than 8 dB/km at 1.090, 1.205, and 1.280 micrometers.

liquid laser A laser whose active medium is in liquid form, such as organic dye and inorganic solutions.

Note: Dye lasers are commercially available; they are often called organic dye or tunable dye lasers.

liquid-phase epitaxy Single-crystal growth technique during which material from a melt precipitates onto a single-crystal substrate of similar lattice constant. The major growth technique for high-quality optoelectronic sources.

lithium niobate Ferroelectric crystalline material of very high electro-optic and piezo-electric coefficients. $LiNbO_3$ is primarily used as substrate and active medium for bulk and thin-film optical modulators and/or switches. The basic material of external modulators for optical communication systems. Other ferroelectric: $LiTaO_3$-lithium tantalate.

Note: For examples see D. Botez and G. Herskowitz, *Proc. IEEE,* vol. 68, pp. 689–732, June 1980.

LNA *Acronym for LAUNCH NUMERICAL APERTURE.*

loaded diffused optical waveguide *See: STRIP-LOADED DIFFUSED OPTI-CAL WAVEGUIDE.*

LOC *Acronym for LARGE OPTICAL-CAV-ITY DIODE.*

log-normal distribution Statistical distribution used for characterizing lifetimes of semiconductor devices or the repair time of electronic systems. The log-normal distribution is a Gaussian distribution of device failures on a logarithmic time scale. After plotting the device failures on log-normal paper a straight line is obtained and two parameters can be determined: the median life and the mean time to failure (MTTF). Used extensively for diode-laser failure characterization.

Note: For examples see W. B. Joyce *et al., Appl. Phys. Lett.,* vol. 28, pp. 684–686, June 1976.

longitudinal balance The electrical symmetry of the two wires of a pair with respect to ground.

longitudinally excited atmosphere laser (LEA) A gas laser in which the electric field excitation of the active medium is longitudinal to (in the direction of) the flow of the active medium.

Note: This type of laser operates in a gas pressure range lower than that required for transverse-excitation.

longitudinal mode One of the ensemble of frequency modes that constitutes the laser spectrum. A longitudinal mode corresponds to one of the oscillation modes that can be allowed along the laser length between the mirror facets.

longitudinal offset loss *See: GAP LOSS.*

long-wavelength light source Diode laser and/or LED emitting in the 1 to 2 μm wavelength range.

loop 1. Go and return conductors of an electric circuit; a closed circuit. 2. A single connection from a switching center or an individual message distribution point, to the terminals of an end instrument. 3. A closed path under measurement in resistance test. 4. A type of antenna used extensively in direction finding equipment. 5. In computer systems, repetition of a group of instructions in a computer routine. 6. In telephone systems, a pair of wires from a central office to the subscriber's telephone. *See also: LINE LOOP; PHASE LOCK LOOP.*

loop-back A method of performing transmission tests of access lines from the serving switching center, that does not require the assistance of personnel at the served terminal. A connection is established over one access line from the serving switching center through the transmission testing and switching equipment and back to the serving switching center over another access line.

loop gain 1. Total usable power gain of a carrier terminal or two-wire repeater. The maximum usable gain is determined by and may not exceed the losses in the closed path. 2. The product of the gain values acting on a signal passing around a closed path loop.

loop noise The noise contribution of the line loops in a system.

loop test A method of testing employed to locate a fault in the insulation of a conductor when the conductor can be arranged to form part of a closed circuit or loop.

loose-tube splicer A glass tube with a square hole used to splice two optical fibers; the curved fibers are made to seek the same corner of the square hole, thus holding them in alignment until the index-matching epoxy, already in the tube, cures, thus forming an aligned, low-loss butted joint. *See also: PRECISION-SLEEVE SPLICER; TANGENTIAL COUPLING.*

LORAN A long-range radionavigation position fixing system using the time difference of reception of pulse type transmissions from two or more fixed stations. (This term is derived from the words "long range electronic navigation".)

LOS *Acronym for LINE OF SIGHT.*

loss *See: ABSORPTION; ANGULAR MISALIGNMENT LOSS; ATTENTUATION; BACKSCATTERING; DIFFERENTIAL MODE ATTENUATION; EXTRINSIC JOINT LOSS; GAP LOSS; INSERTION LOSS; INTRINSIC JOINT LOSS; LATERAL OFFSET LOSS; MACROBEND LOSS; MATERIAL SCATTERING; MICROBEND LOSS; NONLINEAR SCATTERING; RAYLEIGH SCATTERING; REFLECTION; TRANSMISSION LOSS; WAVEGUIDE SCATTERING.*

Lossy medium A wave propagation medium in which a significant amount of the energy of the wave is absorbed per unit distance traveled by the wave, for example, in an optical fiber cladding, lossy material is used to attenuate by absorption the energy that has leaked outside the fiber core.

loupe *See: MAGNIFIER.*

low-loss FEP-clad silica fiber An optical fiber consisting of a pure fused silica core and a Perfluoronated Ethylene-Propylene (FEP) (a commercial polymer) cladding.

Note: FEP-clad silica fibers have refractive indices of 1.458 and 1.338 for the core and cladding respectively, and a transmission loss of 2 to 3 dB/km at the present time, with an ultraviolet capability at 0.546 micrometers of 360 dB/km.

low-loss fiber An optical fiber having a low energy loss, due to all causes, per unit length of fiber, usually measured in dB/km at a specified wavelength.

Note: Low-loss is usually considered to be below 20 dB/km, attenuation in amplitude of a propagation wave is caused primarily by scattering due to metal ions and by absorption due to water in the OH radical form.

low-pass filter　A filter network which passes all frequencies below a specified frequency with little or no loss but which discriminates strongly against higher frequencies.

LP$_{01}$ mode　Designation of the fundamental LP mode. *See: FUNDAMENTAL MODE.*

LPE　*Acronym for LIQUID PHASE EPITAXY.*

LPC　*Acronym for LINEAR PREDICTIVE CODING.*

LP mode　*Abbreviation for LINEARLY POLARIZED MODE.*

lumen　The SI unit of light flux corresponding to $\frac{1}{4\pi}$ of the total light flux emitted by a source having an intensity of 1 candela, thus being equal to the flux issuing from one-sixtieth of a square centimeter of opening of the standard source, and included in a solid angle of one steradian.

lumen-hour　The unit quantity of light equal to one lumen of luminous flux flowing for one hour.

lumen-second　The unit quantity of light equal to one lumen of luminous flux flowing for one second.

lumerg　The centimeter-gram-second unit of luminous energy, equal to 10^{-7} lumen-second.

luminance　The ratio of the luminous intensity emitted by a light source in a given direction by an infinitesimal area of the source, to the projection of that area of the source upon the plane perpendicular to the given direction. *Note:* Luminance is usually stated as luminous intensity per unit area; i.e., luminous flux emitted per unit solid angle projected per unit projected area. The area being the area upon which the flux is incident, or is considered incident, and the area being perpendicular to the direction in which the light wave is propagating.

luminance temperature　The temperature of an ideal blackbody that would have the same luminance as the source for which the luminance temperature is desired for some narrow spectral region.

luminance threshold　*See: ABSOLUTE LUMINANCE THRESHOLD.*

luminescence　The process whereby matter emits electromagnetic radiation which, for certain wavelengths, or restricted regions of the spectrum, is in excess of that attributable to the thermal state of the material and the emissivity of its surface. *Note:* The radiation is characteristic of the particular luminescent material, and occurs without outside stimulation. *See also: PHOSPHORESCENCE.*

luminescent diode　*See: SUPERLUMINESCENT DIODE.*

luminosity　The ratio of luminous flux to the radiant flux in a sample of radiant light flux; for example, lumens per watt of radiant energy. *Synonym: LUMINOUS EFFICIENCY.*

luminosity curve　The curve obtained by plotting luminous relative efficiency against the wavelength of a lightwave. *See also: ABSOLUTE LUMINOSITY CURVE.*

luminous density　The luminous energy per unit volume of an electromagnetic (light) wave.

luminous efficiency　The ratio of the luminous flux emitted to the power consumed by a source of light; for example, lumens per watt applied at the source.

luminous emittance　The total luminous flux emitted by a unit area of an extended surface, in contrast to a point or line source.

luminous flux　The quantity that specifies the capacity of the radiant flux from a light source to produce the attribute of visual sensation known as brightness. *Note:* Luminous flux is radiant flux evaluated with respect to its luminous efficiency of radiation. Unless otherwise stated, luminous flux pertains to the standard photoptic observer.

luminous intensity　The ratio of the luminous flux emitted by a light source, or an element of the source, in an infinitesimally small cone about the given direction, to the solid angle of that cone, usually stated as luminous flux emitted per unit solid angle.

luminous radiation efficiency　*See: LUMINOSITY.*

luminous transmittance　The ratio of the luminous flux transmitted by an object to the incident luminous flux.

lux　A unit of illuminance, equal to a lumen incident per square meter of surface (normal to the direction of propagation).

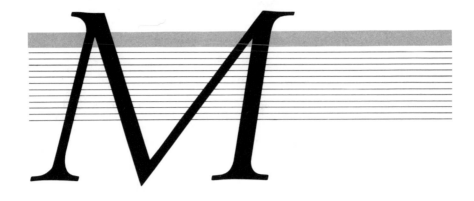

Mach-Zehnder interferometric modulator Guided-wave optics device that produces intensity modulation of guided light. The input signal is equally divided into two waves that are fed into separate waveguides, where each wave can be phase-modulated electro-optically by voltages applied via metallic contacts. When recombined the waves provide an output signal that is intensity-modulated as a function of the phase mismatch between them. Demonstrated mostly on LiNbO$_3$ substrates.

Note: For examples see D. Botez and G. Herskowitz, *Proc. IEEE,* vol. 68, pp. 689–732, June 1980.

macrobending In an optical waveguide, all macroscopic deviations of the axis from a straight line; distinguished from microbending. *See also: MACROBEND LOSS; MICROBEND LOSS; MICROBENDING.*

macrobend loss In an optical waveguide, that loss attributable to macrobending. Macrobending usually causes little or no radiative loss. *Synonym: CURVATURE LOSS. See also: MACROBENDING; MICROBEND LOSS.*

magneto-optic effect The rotation of the plane of polarization of plane-polarized light waves in a medium brought about when subjecting the medium to a magnetic field (Faraday rotation).

Note: The effect can be used to modulate a light beam in a material since many properties such as, conducting velocities, reflection and transmission coefficients at interfaces, acceptance angles, critical angles, and transmission modes, are dependent upon the direction of propagation at interfaces in the media in which the light travels. The amount of rotation is given by:

$$A = VHL$$

where V is a constant, H is the magnetic field strength, and L is the distance. The magnetic field is in the direction of propagation of the light wave. *Synonym: FARADAY EFFECT.*

magnification The ratio of the size of any linear dimension of the image to that of the object in some optical system. *See also: ABSOLUTE MAGNIFICATION; ANGULAR MAGNIFICATION; INDIVIDUAL NORMAL MAGNIFICATION; MAGNIFYING POWER.*

magnifier An optical system, such as a lens or lens system, capable of forming a magnified virtual image of an object placed near its front focal point. *Synonyms: LOUPE; SIMPLE MICROSCOPE; MAGNIFYING GLASS.*

Note: Magnifications of magnifiers range from approximately 3× to 20×.

magnifying glass *See: MAGNIFIER.*

magnifying power The measure of the ability of an optical device to make an object appear larger than it appears to the unaided eye. For example, if an optical element or system has a magnification of 2-power (2×), the object will appear twice as wide and high.

Note: The magnification of an optical instrument is equal to the diameter of the entrance pupil divided by the diameter of the exit pupil. For a telescopic system, the magnification is also equal to the focal length of the eyepiece. Another expression for the magnification of an instrument is the tangent of an angle in the apparent field divided by the tangent of the corresponding angle in the true field. *Synonym: MAGNIFICATION.*

major In optics, a blank to which a piece of glass of a different index of refraction will be fused to make a multifocal lens.

mangin mirror A negative meniscus lens whose second, or convex, surface is silvered.

Note: Spherical aberration can be corrected for any given position of the image by carefully choosing the radii.

mark In binary communications, one of the two significant conditions of modulation. *Synonyms: MARKING PULSE; MARKING SIGNAL. See also: MARKING BIAS; SIGNAL TRANSITION; SIGNIFICANT CONDITION OF MODULATION; SPACE.*

marking bias The uniform lengthening of all marking signal pulses at the expense of all spacing pulses. *See also: BIAS; MARK; SPACING BIAS.*

marking pulse *Synonym: MARK.*

marking signal *Synonym: MARK.*

maser *See: MICROWAVE AMPLIFICATION BY STIMULATED EMISSION OF RADIATION; OPTICAL MASER.*

master clock A clock that generates accurate timing signals for the control of other clocks and possibly other equipment. *See also: REFERENCE CLOCK.*

master frequency generators In FDM, equipments used to provide system end-to-end carrier frequency synchronization and frequency accuracy of tones transmitted over the system.
Note: The following types of oscillators are employed in the DCS FDM systems: Type 1. Master carrier oscillator as an integral part of the multiplexer set. Type 2. Submaster oscillator equipment or slave oscillator equipment as an integral part of the multiplex set. Type 3. External master oscillator equipment having extremely accurate and stable characteristics. *Synonym: MASTER FREQUENCY OSCILLATORS.*

master frequency oscillators *Synonym: MASTER FREQUENCY GENERATORS.*

mastergroup *See: GROUP.*

master-slave timing A system wherein one station or node supplies the timing reference for all other interconnected stations or nodes. *See also: TIMING SIGNAL.*

master station In a data network, the station that has been requested by the control station to ensure data transfer to one or more slave stations.
Note: A master station has control of one or more data links of the data communication network at a given instant. The assignment of master status to a given station is temporary and is controlled by the control station according to the procedures set forth in the operational protocol. Master status is normally con-

ferred upon a station so that it may transmit a message, but a station need not have a message to send to be nominated as master. *See also: CONTENTION; CONTROL STATION; DATA COMMUNICATION; DATA TRANSMISSION; INTERROGATION; NETWORK; PRIMARY STATION; SECONDARY STATION; SLAVE STATION; TRIBUTARY STATION.*

matching materials *See: INDEX-MATCHING MATERIALS.*

material *See: ISOTROPIC MATERIAL.*

material absorption *See: ABSORPTION.*

material dispersion That dispersion attributable to the wavelength dependence of the refractive index of material used to form the waveguide. Material dispersion is characterized by the material dispersion parameter M. *See also: DISPERSION; DISTORTION; MATERIAL DISPERSION PARAMETER; PROFILE DISPERSION PARAMETER; WAVEGUIDE DISPERSION.*

material dispersion parameter (M)

$$M(\lambda) = -1/c \, (dN/d\lambda)$$
$$= \lambda/c \, (d^2n/d\lambda^2)$$

where n is the refractive index,
N is the group index: $N = n - \lambda(dn/d\lambda)$,
λ is the wavelength, and
c is the velocity of light in vacuum.
Note 1. For many optical waveguide materials, M is zero at a specific wavelength λ_0, usually found in the 1.2 to 1.5μm range. The sign convention is such that M is positive for wavelengths shorter than λ_0 and negative for wavelengths longer than λ_0.
Note 2. Pulse broadening caused by material dispersion in a unit length of optical fiber is given by M times spectral linewidth ($\Delta\lambda$), except at $\lambda = \lambda_0$, where terms proportional to $(\Delta\lambda)^2$ are important. (See Note 1.) *See also: GROUP INDEX; MATERIAL DISPERSION.*

material scattering In an optical waveguide, that part of the total scattering attributable to the properties of the materials used for waveguide fabrication. *See also: RAYLEIGH SCATTERING; SCATTERING; WAVEGUIDE SCATTERING.*

materials Index-matching materials.

maximal-ratio combiner A diversity combiner in which the signals from each channel are added together; the gain of each channel is made proportional to the rms signal and inversely proportional to the mean square noise in that channel, with the same proportionality constant for all channels. *Synonym: RATIO-SQUARED COMBINER. See also: DIVERSITY COMBINER; SELECTIVE COMBINER.*

maximum acceptance angle The maximum angle between the longitudinal axis of an optical transmission medium such as an optical fiber or a deposited optical film, and the normal to the wave front, i.e., the direction of the entering light ray, in order that these be total internal reflection of the portion of incident light, that is transmitted through the fiber interface, i.e., the angle between the transmitted ray and the normal to the inside surface of the cladding is greater than the critical angle. *Note:* The maximum acceptance angle is given by the square root of the difference of the squares of the indices of refraction of the fiber core glass and the cladding. The square root of the difference of the squares is called the numerical acceptance (NA).

MBE *Acronym for MOLECULAR BEAM EPITAXY.*

MCSP laser *See: MODIFIED CHANNELED-SUBSTRATE-PLANAR DIODE LASER.*

MCVD *Acronym for MODIFIED CHEMICAL-VAPOR DEPOSITION PROCESS.*

mean power The power supplied to the antenna transmission line by a radio transmitter during normal operation, averaged over a time sufficiently long compared with the period of the lowest frequency encountered in the modulation. A time of $\frac{1}{10}$ second during which the mean power is greatest will be selected normally.

mean spherical intensity The average value of intensity of an electromagnetic source of radiation such as a light source, with respect to all directions.

mean time to failure (MTTF) A statistical measure of device lifetime. The product of median life and exp $(\sigma^2/2)$ where σ is the standard deviation of a log-normal distribution. *See: LOG-NORMAL DISTRIBUTION; MEDIAN LIFE.*

measure *See: LENS MEASURE.*

measurement period *See: PERFORMANCE MEASUREMENT PERIOD.*

mechanical splice A fiber splice accomplished by fixtures or materials, rather than by thermal fusion. Index matching material may be applied between the two fiber ends. *See also: FUSION SPLICE; INDEX MATCHING MATERIAL; OPTICAL WAVEGUIDE SPLICE.*

median life In device lifetesting the time at which half of the device population fails to meet a preset specification. For instance for diode lasers end of life is sometimes defined as the time at which the drive current has to be increased by 50% to maintain a given output power level.

medium In optics, any substance or space through which light can travel. *See also: LOSSY MEDIUM.*

medium-loss fiber An optical fiber having a medium-level energy loss, due to all causes per unit length of fiber, usually measured in dB/km at a specified wavelength. *Note:* Medium-loss is usually considered to be between 20 and 100 dB/km. Attenuation in amplitude of a propagating wave is caused primarily by scattering due to metal ions and by absorption due to water in the OH radical form.

meniscus A lens having a convex and a concave surface. *Synonym: CONCAVO-CONVEX LENS. See also: DIVERGENT MENISCUS LENS.*

meridian plane Any plane that contains the optical axis of an optical system, such as a plane that contains the optical axis of a round optical fiber.

meridional ray A ray that passes through the optical axis of an optical waveguide (in contrast with a skew ray, which does not.) *See also: AXIAL RAY; GEOMETRIC OPTICS; NUMERICAL APERTURE; OPTICAL AXIS; PARAXIAL RAY; SKEW RAY.*

mesochronous The relationship between two signals such that their corresponding significant instants occur at the same average rate. *See also: ANISOCHRONOUS; HETEROCHRONOUS; HOMOCHRONOUS; ISOCHRONOUS; PLESIOCHRONOUS; SYNCHRONOUS SYSTEM.*

message switching A method of handling message traffic through a switching center, either from local users or from other switching centers, whereby a connection is established between the calling and called stations or the message traffic is stored and forwarded through the system. *See also: CIRCUIT SWITCHING; PACKET SWITCHING; QUEUE TRAFFIC; STORE-AND-FORWARD*

SWITCHING CENTER; SWITCHING SYSTEM.

metalorganic chemical vapor deposition *See: ORGANO-METALLIC VAPORPHASE EPITAXY.*

meter *See: PHOTOCONDUCTIVE METER; PHOTOVOLTAIC METER.*

microbending loss In an optical fiber, the loss or attenuation in signal power caused by small bends, kinks, or abrupt discontinuities in direction of the fibers, usually caused by fiber cabling or by wrapping fibers on drums.

Note: Microbending losses usually result from a coupling of guided modes among themselves and among the radiation modes.

microbend loss In an optical waveguide, that loss attributable to microbending. *See also: MACROBEND LOSS.*

micrometer A unit of length in the metric system equal to 0.001 millimeter or 0.000001 meter, i.e., one millionth of a meter.

microscope *See: MAGNIFIER.*

microwave A term loosely applied to those radio frequency wavelengths which are sufficiently short to exhibit some of the properties of light; e.g., they are easily concentrated into a beam. Commonly used for frequencies from about 1 GHz to 30 GHz.

microwave amplification by stimulated emission of radiation (maser) A low-noise radio-frequency amplifier whose emission energy is stored in a molecular or atomic system by a microwave power supply stimulated by the input signal.

middle infrared Pertaining to electromagnetic wavelengths from 3 to 30 micrometers.

millimicron A unit of length in the metric system equal to 0.001 micrometer, or 10 angstroms.

minus lens *See: DIVERGING LENS.*

mirror 1. A flat surface optically ground and polished on a reflecting material, or a transparent material that is coated to make it reflecting, used for reflecting light.

Note: A beam splitting mirror has a lightly deposited metallic coating that transmits a portion of the incident light and reflects the remainder.

2. A smooth highly polished plane or curved surface for reflecting light.

Note: Usually a thin coating of silver or aluminum on glass constitutes the actual reflecting surface. When this surface is applied to the front face of the glass, the mirror is a front-surface mirror. *See also: BACK-SURFACE MIRROR; FRONT-SURFACE MIRROR; MANGIN MIRROR; OFF-AXIS PARABOLOIDAL MIRROR; PARABOLOIDAL MIRROR; TRIPLE MIRROR.*

mirror reflectivity The (power) reflection coefficient at a mirror facet for a given optical mode in a diode laser or for the light in an LED. For uncoated diodes the reflection coefficient at one facet is approximately 30%. Quarter-wave coatings can bring the mirror reflectivity down to 1%, while dichroic reflectors or metal coatings can increase the reflectivity to 95%.

Note: For examples see M. Ettenberg, *Appl. Phys. Lett.,* vol. 32, pp. 724–726, 1976.

misalignment loss *See: ANGULAR MISALIGNMENT LOSS; GAP LOSS; LATERAL OFFSET LOSS.*

mismatch loss *See: REFRACTIVE-INDEX-PROFILE MISMATCH LOSS.*

mismatch-of-core-radii loss A loss of signal power introduced by an optical fiber splice in which the radii of the cores of the two fibers that are joined are not equal.

Note: The loss if usually expressed in decibels (dB).

MIVPO *Acronym for MODIFIED INSIDE-VAPOR PHASE-OXIDATION PROCESS.*

mixed-gas laser An ion laser that uses a mixture of gases, such as argon and krypton, as the active medium.

MO-CVD *Acronym for METALORGANIC CHEMICAL VAPOR DEPOSITION.*

modal gain In a semiconductor diode laser the gain for a given optical mode, usually the fundamental one. At lasing threshold the modal gain equals the internal cavity losses and the mirror transmission losses. Symbol: G.

modal loss In an open waveguide, such as an optical fiber, a loss of energy on the part of an electromagnetic wave due to obstacles outside the waveguide, abrupt changes in direction of the waveguide, or other anomalies, that cause changes in the propagation mode of the wave in the waveguide. *See also: PROPAGATION MODE.*

modal noise Noise generated in an optical fiber system by the combination of mode dependent optical losses and fluc-

tuation in the distribution of optical energy among the guided modes or in the relative phases of the guided modes. *Synonym: SPECKLE NOISE. See also: MODE.*
Note: For example see R.E. Epworth, *Laser Focus,* pp. 109–113, Sept. 1981.

mode In any cavity or transmission line, one of those electromagnetic field distributions that satisfies Maxwell's equations and the boundary conditions. The field pattern of a mode depends on the wavelength, refractive index, and cavity or waveguide geometry. *See also: BOUND MODE; CLADDING MODE; DIFFERENTIAL MODE ATTENUATION; DIFFERENTIAL MODE DELAY; EQUILIBRIUM MODE DISTRIBUTION; EQUILIBRIUM MODE SIMULATOR; FUNDAMENTAL MODE; HYBRID MODE; LEAKY MODES; LINEARLY POLARIZED MODE; MODE VOLUME; MULTIMODE DISTORTION; MULTIMODE LASER; MULTIMODE OPTICAL WAVEGUIDE; SINGLE MODE OPTICAL WAVEGUIDE; TRANSVERSE ELECTRIC MODE; TRANSVERSE MAGNETIC MODE; UNBOUND MODE.*

mode coupling In an optical waveguide, the exchange of power among modes. The exchange of power may reach statistical equilibrium after propagation over a finite distance that is designated the equilibrium length. *See also: EQUILIBRIUM LENGTH; EQUILIBRIUM MODE DISTRIBUTION; MODE; MODE SCRAMBLER.*

mode dispersion Often erroneously used as a synonym for *MULTIMODE DISTORTION.*

mode fiber *See: SINGLE-MODE FIBER.*

mode filter A device used to select, reject, or attenuate a certain mode or modes.

MODEM *Acronym for MODULATOR-DEMODULATOR.* A device that modulates and demodulates signals.
Note: 1. Modems are primarily used for converting digital signals into quasi-analog signals for transmission and for reconverting the quasi-analog signals into digital signals.
Note: 2. Many additional functions may be added to a modem to provide for customer service and control features. *Synonym: SIGNAL CONVERSION EQUIPMENT. See also: DATA CIRCUIT-TERMINATING EQUIPMENT; INPUT/OUTPUT DEVICE; NARROWBAND MODEM; QUASI-ANALOG SIGNAL; WIDEBAND MODEM.*

mode mixer *Synonym: MODE SCRAMBLER.*

mode number In diode lasers a means of enumerating the spatial modes. The fundamental mode has the mode number 0, and the high-order modes have numbers from 1 up. Symbol: m.
Note: The first-order mode has $m = 1$ and is the first high-order mode (i.e., two maxima and a null).

mode (or modal) distortion *Synonym: MULTIMODE DISTORTION.*

mode-partition noise Noise that occurs in single-mode fiber systems after the propagation of light from a multi-longitudinal-mode source. During signal propagation through long lengths of single-mode fiber, the material wavelength dispersion separates the longitudinal modes, reducing the correlation of individual modes fluctuations, and thus causing excess noise.

modes *See: COUPLED MODES.*

mode scrambler 1. A device for inducing mode coupling in an optical fiber. 2. A device composed of one or more optical fibers in which strong mode coupling occurs.
Note: Frequently used to provide a mode distribution that is independent of source characteristics or that meets other specifications. *Synonym: MODE MIXER. See also: MODE COUPLING.*

mode-stabilized laser Diode laser for which the spatial mode pattern, usually the fundamental mode, is stabilized as a function of drive current above threshold. Spatial-mode stabilization is mostly realized by the creation of a laterally waveconfining structure (waveguide or antiguide).

mode stripper *See: CLADDING MODE STRIPPER.*

mode volume The number of bound modes that an optical waveguide is capable of supporting; for $V > 5$, approximately given by $V^2/2$ and $(V^2/2)[g/(g + 2)]$, respectively, for step index and power-law profile waveguides, where g is the profile parameter, and V is normalized frequency. *See also: EFFECTIVE MODE VOLUME; MODE; NORMALIZED FREQUENCY; POWER-LAW INDEX PROFILE; STEP INDEX PROFILE; V NUMBER.*

modified AMI An AMI signal that does

not strictly conform with alternate mark inversion but includes violations in accordance with a defined set of rules. *See also: ALTERNATE MARK INVERSION SIGNAL; SIGNAL.*

modified channel substrate planar (MCSP) diode laser Mode-stabilized diode laser structure grown by one-step liquid-phase epitaxy over a V-shaped channel in an absorbing substrate, and for which the lasing lateral spot size is limited to values below 2.5 μm (full width at $1/e^2$ points intensity). Typical reliable power capability: 3–5 mW/facet CW. A low-power version of the CSP structure, intended for use in digital audio-disk readout. Similar to the CNS laser except for the use of very thin (500–1000Å) active layers. Symbol: MCSP.

Note: See H. Matsueda and M. Nakamura, *Tech. Digest Int. Electron. Dev. Meeting*, Paper 12.1, pp. 272–275, Washington, DC, December 1981.

modified chemical vapor deposition process (MCVD) A modified inside vapor phase oxidation process for production of optical fibers in which the burner travels along the glass tube and the soot particles are created inside the tubing rather than in the burner flame as in the OVPO process.

Note: The chemical reactants, such as silicon tetrachloride, oxygen, and dopants, are caused to flow through the rotating tube of glass at a pressure of about one atmosphere, the high temperature causing the formation of oxides (soot) and a glassy deposit on the inside tube surface, and the tube then being drawn into a solid fiber. *Synonym: MODIFIED INSIDE VAPOR PHASE OXIDATION PROCESS. See also: CHEMICAL VAPOR PHASE OXIDATION PROCESS.*

modified inside vapor phase oxidation process (MIVPO) *See: MODIFIED CHEMICAL VAPOR DEPOSITION PROCESS.*

modulation A controlled variation with time of any property of a wave for the purpose of transferring information.

modulation factor In amplitude modulation, the ratio of the peak variation actually used to the maximum design variation in a given type of modulation.

Note: In conventional amplitude modulation the maximum design variation is considered that for which the instanta-

neous amplitude of the modulated signal reaches zero. When zero is reached, the modulation is considered 100 percent. *See also: MODULATION INDEX; UNBALANCED MODULATOR.*

modulation index In angle modulation, the ratio of the frequency deviation of the modulated signal to the frequency of a sinusoidal modulating signal. *See also: MODULATION FACTOR.*

modulation rate The reciprocal of the unit interval measured in seconds. This rate is expressed in baud.

modulation transfer function In optics, the function, usually a graph, describing the modulation of the image of a sinusoidal object as the frequency increases. *Synonym: SINE WAVE RESPONSE, CONTRAST TRANSFER FUNCTION.*

modulator *See: INTEGRATED-OPTICAL CIRCUIT FILTER-COUPLER-SWITCH-MODULATOR; THIN-FILM OPTICAL MODULATOR.*

molecular beam epitaxy Single-crystal growth technique during which atoms from several converging particle beams are deposited onto a single-crystal substrate of similar lattice constant. MBE produces crystalline layers of very uniform thickness and composition. A tool for studying quantum effects in some optoelectronic and microwave devices.

molecular laser A type of gas laser whose active medium is a molecular substance (compound), for example, a carbon dioxide, hydrogen cyanide, or water vapor laser.

molecular stuffing process (MS) A process of making graded refractive index optical fibers using five broad steps, namely, glass melting, phase separation, leaching, dopant introduction, and consolidation.

monochromatic Consisting of a single wavelength or color. In practice, radiation is never perfectly monochromatic but, at best, displays a narrow band of wavelengths. *See also: COHERENT; LINE SOURCE; SPECTRAL WIDTH.*

monochromatic light Electromagnetic radiation, in the visible or near visible (light) portion of the spectrum, that has only one frequency or wavelength.

monochromatic radiation Electromagnetic radiation that has one frequency or wavelength. *See also: POLYCHROMATIC RADIATION.*

monochromator An instrument for isolating narrow portions of the spectrum.

monomode optical waveguide *Synonym: SINGLE MODE OPTICAL WAVEGUIDE.*

mounting cement An adhesive used to hold optical elements in their mounts. It may be either a thermoplastic, thermosetting, or chemical-hardening material.

MS *Acronym for MOLECULAR STUFFING PROCESS.*

MTTF *Acronym for MEAN TIME TO FAILURE.*

multichannel bundle cable In optical fiber systems, two or more single-bundle cables all in one outside jacket.

multichannel cable In optical fiber systems, two or more cables combined in a single jacket, harness, strength-member, cover, or other unitizing element.

multichannel single fiber cable In optical fiber systems, two or more single-fiber cables all in one outside jacket.

multifiber cable An optical cable that contains two or more fibers, each of which provides a separate information channel. *See also: FIBER BUNDLE; OPTICAL CABLE ASSEMBLY.*

multifiber joint An optical splice or connector designed to mate two multifiber cables, providing simultaneous optical alignment of all individual waveguides.
Note: Optical coupling between aligned waveguides may be achieved by various techniques including proximity butting (with or without index matching materials), and the use of lenses.

multifocal In optics, pertaining to a system or component, such as a lens or lens system, that has, or is characterized by, two or more foci.

multifrequency pulsing *Synonym: MULTIFREQUENCY SIGNALING.*

multifrequency signaling A signaling method using combinations of two-out-of-six (MF 2/6) or two-out-of-eight (MF 2/8) voice-band frequencies to indicate telephone address digits, precedence ranks, and line or trunk busy. "MF 2/6" uses frequencies of 700, 900, 1100, 1300, 1500, and 1700 Hz; "MF 2/8" uses 697, 770, 852, 941, 1209, 1336, 1447, and 1633 Hz. *Synonym: MULTIFREQUENCY PULSING.*

multilayer filter *See: INTERFERENCE FILTER.*

multi-level modulation *Synonym: M-ARY SIGNALING.*

multiline laser A multimode gas laser.

multimode dispersion *See: OPTICAL MULTIMODE DISPERSION.*

multimode distortion In an optical waveguide, that distortion resulting from differential mode delay.
Note: The term "multimode dispersion" is often used as a synonym; such usage, however, is erroneous since the mechanism is not dispersive in nature. *Synonyms: INTERMODAL DISTORTION; MODE (OR MODAL) DISTORTION. See also: DISTORTION.*

multimode fiber An optical fiber waveguide that will allow more than one mode to propagate.
Note: Optical fibers have a much larger core (25–75 micrometers) than singlemode fibers (2–8 micrometers) and thus permit nonaxial rays or modes to propagate through the core compared with only one mode through a single-mode fiber.

multimode group delay *Synonym: DIFFERENTIAL MODE DELAY.*

multimode group-delay spread In an optical waveguide, such as an optical fiber, slab dielectric waveguide, or integrated optical circuit, the variation in group delay, due to differences in group velocity, among the supported propagating modes at a single frequency.
Note: Multimode group-delay spread contributes to group-delay distortion, along with material dispersion and waveguide-delay distortion. The spread in arrival time of the edges of a light pulse at the end of an optical waveguide, such as an optical fiber or bundle, is caused by the different time delays, or propagation times, of the various waveguide modes. The modes can be visualized as different optical paths of different lengths, for example, photons or waves propagating along the fiber axis reach the end of the fiber before photons or waves that propagate along semihelical off-axis paths, causing optical pulse spreading (increase in width) and resulting intersymbol interference beyond the bit-rate-length capacity of the fiber. Reduction of spreading can be accomplished by having the longer paths be in a lower-refractive-index medium so that the wave can travel faster in the longer paths.

multimode laser A laser that produces emission in two or more transverse or longitudinal modes. *See also: LASER; MODE.*

multimode optical waveguide An optical waveguide that will allow more than one bound mode to propagate.
Note: May be either a graded index or step index waveguide. *See also: BOUND MODE; MODE; MODE VOLUME; MULTIMODE DISTORTION; NORMALIZED FREQUENCY; POWER-LAW INDEX PROFILE; SINGLE MODE OPTICAL WAVEGUIDE; STEP INDEX OPTICAL WAVEGUIDE.*

multimode waveguide A waveguide capable of supporting more than one electromagnetic propagation mode.

multipath The propagation phenomenon that results in radio signals reaching the receiving antenna by two or more paths.
Note: Multipath effects range from constructive reinforcement to destructive cancellation of the signal. *See also: PROPAGATION; RAYLEIGH FADING.*

multiple access 1. The connection of a user or subscriber to two or more switching centers by separate access lines using a single message routing indicator or telephone number. 2. In satellite communications, the capability of a communications satellite to function as a portion of a communications link between more than one pair of satellite terminals simultaneously.
Note: Three types of multiple access are presently employed with communications satellites: code division, frequency division, and time division. *See also: DUAL ACCESS.*

multiple-bundle cable A number of jacketed optical fiber bundles placed together in a common, usually cylindrical, envelope.

multiple-bundle cable assembly A multiple bundle optical fiber cable terminated and ready for installation.

multiple-fiber cable Two or more jacketed fibers placed together in a common envelope.

multiple-fiber cable assembly A multiple fiber cable terminated and ready for installation.

multiple frequency shift keying (MFSK) A form of frequency-shift keying in which multiple codes are used in the transmission of digital signals.

Note: The coding systems may utilize multiple frequencies transmitted concurrently or sequentially.

multiple homing 1. In telephony, the connection of a terminal facility so that it can be served by one of several switching centers.
Note: This service may use a single directory number.
2. In telephony, the connection of a terminal facility to more than one switching center by separate access lines. Separate directory numbers are applicable to each accessed switching center. *See also: DUAL HOMING.*

multiplex aggregate bit rate The bit rate in a time division multiplexer that is equal to the sum of the input channel data signaling rates available to the user plus the rate of the overhead bits required.
Note: Mathematically, the multiplex aggregate bit rate, MABR, is given by:

$$MABR = R(\sum_{i=1}^{m} n_i + H)$$

where n_i is the number of bits per multiplex frame associated with the i-th input channel,

m is the maximum number of input channels to the multiplexer (including non-working channels, or equipped channels, or both),

H is the number of overhead bits per multiplex frame of the output channel, and

R is the repetition rate of the frame of the output channel.

The number of bits in the multiplex frame is assumed to be constant. *See also: TIME-DIVISION MULTIPLEX.*

multiplex baseband The frequency band occupied by the aggregate of the signals in the line interconnecting the multiplexing and radio or line equipments.

multiplex baseband receive terminals The point in the baseband circuit nearest the multiplex equipment from which connection is normally made to the radio baseband receive terminals or intermediate facility. *See also: BASEBAND.*

multiplex baseband send terminals The point in the baseband circuit nearest the multiplex equipment from which connection is normally made to the radio baseband send terminals or intermediate facility. *See also: BASE-BAND.*

multiplex (MUX) Use of a common channel to make two or more channels, either by splitting of the frequency band transmitted by the common channel into narrower bands, each of which is used to constitute a distinct channel (frequency division multiplex), or by allotting this common channel to multiple users in turn, to constitute different intermittent channels (time division multiplex). *See also: ADAPTIVE CHANNEL ALLOCATION; ASYNCHRONOUS TIME DIVISION MULTIPLEXING; BIT STUFFING; DEMULTIPLEX; DIGITAL BLOCK; FREQUENCY DIVISION MULTIPLEX; WAVELENGTH DIVISION MULTIPLEX; HETEROGENEOUS MULTIPLEXING; HOMOGENEOUS MULTIPLEXING; ORDER-WIRE MULTIPLEX; ORTHOGONAL MULTIPLEX; PCM MULTIPLEX EQUIPMENT; STATISTICAL MULTIPLEXING; STATISTICAL TIME-DIVISION MULTIPLEXING; TIME-DIVISION MULTIPLEX; TRUNK GROUP MULTIPLEXER (TGM).*

multiplexer *See: THIN-FILM OPTICAL MULTIPLEXER.*

multiplexer-filter *See: FIBER-OPTIC ROD MULTIPLEXER-FILTER.*

multiplexing *See: WAVELENGTH DIVISION MULTIPLEXING (WDM).*

multiplication factor *See: AVALANCHE GAIN.*

multipoint circuit A circuit providing simultaneous transmission among three or more separate points.

multipoint link A data communication link connecting two or more stations.

multiport coupler *See: FIBER-OPTIC MULTIPORT COUPLER.*

multi-refracting crystal A transparent crystalline substance that is anisotropic with respect to its refractive index in different directions, i.e. to the velocity of light traveling within it in different directions.

mutually synchronized network A network synchronizing arrangement in which each clock in the network exerts a degree of control on all others.

MUX *Acronym for MULTIPLEX.*

NA *Acronym for NUMERICAL APER-TURE.*

NAM laser *See: NONABSORBING-MIR-ROR LASER.*

narrowband modem A modem whose modulated output signal has an essential frequency spectrum that is limited to that which can be wholly contained within, and faitffully transmitted through, a voice channel with a nominal 4 kHz bandwidth. *See also: MODEM; WIDEBAND MODEM.*

narrowband signal Any analog signal or analog representation of a digital signal whose essential spectral content is limited to that which can be contained within a voice channel of nominal 4 kHz bandwidth.

narrow-stripe laser Spatially mode-stabilized diode laser of planar DH geometry and for which the current is confined to stripe contacts 2 to 7 μm-wide. NS lasers have wide, non-Gaussian astigmatic beams, and oscillate in a multi-longitudinal-mode spectrum. *See: GAIN-GUIDED DIODE LASER.*
Note: Sometimes referred as "multimode" lasers to be used in multimode fiber systems for modal noise minimization.

N-ary digital signals Digital signals in which a signal element may assume N discrete states. *See also: M-ARY CODE; M-ARY SIGNALING; N-ARY INFORMATION ELEMENT.*

N-ary information element An information element enabling the representation of N distinct states. *See also: M-ARY CODE; M-ARY SIGNALING; N-ARY DIGITAL SIGNALS.*

near-end crosstalk Crosstalk which is propagated in a disturbed channel in the direction opposite to the direction of propagation of the current in the disturbing channel. The terminals of the disturbed channel, at which the near-end crosstalk is present, and the energized terminal of the disturbing channel, are usually near each other.

near field *Synonym: NEAR-FIELD REGION.*

near-field diffraction pattern The diffraction pattern observed close to a source or aperture, as distinguished from far-field diffraction pattern.
Note: The pattern in the output plane of a fiber is called the near-field radiation pattern. *Synonym: FRESNEL DIFFRACTION PATTERN. See also: DIFFRACTION; FAR-FIELD DIFFRACTION PATTERN; FAR-FIELD REGION.*

near-field pattern *Synonym: NEAR-FIELD RADIATION PATTERN. See: RADIATION PATTERN.*

near-field radiation pattern *See: RADIATION PATTERN.*

near-field region The region close to a source, or aperture. The diffraction pattern in this region typically differs significantly from that observed at infinity and varies with distance from the source. *See also: FAR-FIELD DIFFRACTION PATTERN; FAR-FIELD REGION.*

near-field scanning The technique for measuring the index profile of an optical fiber by illuminating the entrance face with an extended source and measuring the point-by-point radiance of the exit face. *See also: REFRACTED RAY METHOD.*

near infrared Pertaining to electromagnetic wavelengths from 0.75 to 3 micrometers.

near zone *Synonym: NEAR-FIELD REGION.*

negative-index guide A waveconfining dielectric structure for which the index of refraction of the central part is below the indices of refraction of the outer parts. *See: ANTIGUIDE.*

negative lens *See: DIVERGING LENS.*

negative meniscus *See: DIVERGENT MENISCUS LENS.*

negative photodiode coupler *See: POSITIVE-INTRINSIC-NEGATIVE PHOTODIODE COUPLER.*

negative pulse stuffing *Synonym: DE-STUFFING.*

NEP *Acronym for NOISE EQUIVALENT POWER.*

neper A standard unit for expressing transmission gain or loss and relative ratios. Like the decibel, it is a dimensionless unit, and ITU (CCITT and CCIR) Recommendations recognize both units. It is a logarithmic unit, based on natural logarithms instead of common logarithms. The neper is equal to the natural logarithm of the scalar ratio of two currents or two voltages, such that nepers, N, are given by:

$$N = \ln (I_1/I_2) = \ln (E_1/E_2)$$

Note: One neper (N) = 8.686 dB. The units decineper (N/10) and centineper (N/100) are also used. The ln is the natural logarithm, i.e., the logarithm to the base e, where e equals about 2.71828. The 8.686 is from 20/ln 10. *See also: dBm.*

Newton's fringes *See: NEWTON'S RINGS.*

Newton's rings The series of rings, bands, or fringes formed when two clean polished surfaces are placed in contact with a thin air film between them and reflected, usually chromatic, beams of light from the two adjacent surfaces interfere with each other, causing alternate cancellation and reinforcement of light as the distance between the surfaces are multiples or submultiples of the wavelength.

Note: By counting these bands from the point of actual contact the departure of one surface from the other is determined. The regularity of the fringes maps out the regularity of the distance between the two surfaces. This is the usual method of determining the fit of a surface under test to a standard surface of a test glass. *Synonym: NEWTON'S FRINGES.*

NF *Acronym for NOISE FIGURE.*

NMI *Acronym for NAUTICAL MILE.*

noise 1. An undesired disturbance within the useful frequency band. 2. The summation of unwanted or disturbing energy introduced into a communications system from man-made and natural sources. *See also: ACOUSTIC NOISE; AMBIENT NOISE LEVEL; ANTENNA GAIN-TO-NOISE TEMPERATURE; ANTENNA NOISE TEMPERATURE; ATMOSPHERIC NOISE; BACKGROUND NOISE; BLACK NOISE; BLUE NOISE; CARRIER NOISE LEVEL; CHANNEL NOISE LEVEL; CIRCUIT NOISE LEVEL; COSMIC NOISE; EFFECTIVE INPUT NOISE TEMPERATURE; EQUIPMENT INTERMODULATION NOISE; EQUIVALENT NOISE RESISTANCE; EQUIVALENT PCM NOISE; FEEDER ECHO NOISE; FRONT-END NOISE TEMPERATURE; IDLE-CHANNEL NOISE; IMPULSE NOISE; INBAND NOISE POWER RATIO; INTERMODULATION NOISE; INTRINSIC NOISE; LOOP NOISE; NOTCHED NOISE; PATH INTERMODULATION NOISE; PRECIPITATION STATIC; PSEUDORANDOM NOISE; QUANTIZING NOISE; RANDOM NOISE; RECEIVED NOISE POWER; REFERENCE NOISE; SHOT NOISE; SIGNAL-PLUS-NOISE TO NOISE RATIO; SIGNAL-TO-NOISE RATIO; SINGLE-SIDEBAND NOISE POWER RATIO; THERMAL NOISE; TOTAL CHANNEL NOISE; WHITE NOISE.*

noise equivalent power (NEP) At a given modulation frequency, wavelength, and for a given effective noise bandwidth, the radiant power that produces a signal-to-noise ratio of 1 at the output of a given detector.

Note: 1. Some manufacturers and authors define NEP as the minimum detectable power per root unit bandwidth; when defined in this way, NEP has the units of watts/(hertz)$^{1/2}$. Therefore, the term is a misnomer, because the units of power are watts. *See also: D*; DETECTIVITY.*

Note: 2. Some manufacturers define NEP as the radiant power that produces a signal-to-dark-current noise ratio of unity. This is misleading when dark-current noise does not dominate, as is often true in fiber systems.

noise factor *Synonym: NOISE FIGURE.*

noise figure (NF) The ratio of the output noise power of a device to the portion thereof attributable to thermal noise in the input termination at standard noise temperature (usually 290 K).

Note: The noise figure is thus the ratio of actual output noise to that which would remain if the device itself were made noiseless. In heterodyne systems, output noise power includes spurious contributions from image-frequency transformations, but the portion attributable to thermal noise in the input termination at standard noise temperature includes only that which appears in the output via the principal frequency transformation of the system and excludes that which appears via the image frequency transformation. *Synonym: NOISE FACTOR.*

noise level The volume of noise power, measured in dB, referred to a base.

noise-limited operation *See: DETECTOR NOISE-LIMITED OPERATION; THERMAL NOISE-LIMITED OPERATION.*

noise measurement units Noise is usually measured in terms of power, either relative or absolute. The decibel is the unit for most of these measurements, although the picowatt is also used. A suffix is usually added to denote a particular reference base or specific qualities of the measurement. Examples of noise measurement units are dBa, dBa(F1A), dBa(HA1), dBa0, dBm, dBm0, dBm(Psoph), dBm0p, dBrn, dBrn(144-line), dBrnC, dBrn(f_1–f_2), pW, pWp, and pWp0. *See also: NOISE WEIGHTING.*

noise power The mean power supplied to the antenna transmission line by a radio transmitter when loaded with white noise having a Gaussian amplitude distribution.

noise suppression A receiver circuit arrangement that automatically reduces the noise output during periods when no carrier is being received. *See also: SQUELCH CIRCUIT.*

noise weighting A specific amplitude-frequency characteristic which permits a measuring set to give numerical readings which approximate the interfering effects to any listener using a particular class of telephone instrument.
Note: 1. Noise weighting measurements are made in lines terminated either by the measuring set or the class of instrument.
Note: 2. The noise weightings generally used were established by agencies concerned with public telephone service, and are based on characteristics of specific commercial telephone instruments, representing successive stages of technological development. The coding of com-

mercial apparatus appears in the nomenclature of certain weightings. The same weighting nomenclature and units are used in military versions of commercial noise measuring sets. *See also: dBa, dBrn ADJUSTED; dBmOp; dBrn; dBrn(f_1–f_2); NOISE MEASUREMENT UNITS; PSOPHOMETRIC VOLTAGE; pWp; WEIGHTING NETWORK.*

nonabsorbing-mirror laser Diode laser, especially of the AlGaAs type, for which the facet-mirror regions are made transparent to the lasing light. Then catastrophic diode damage is not initiated at the facets and thus the power capabilities increase five to tenfold. *See: CRANK-TJS DIODE LASER; "WINDOW-STRIPE" Zn-DIFFUSED DIODE LASER.*

noncoherent bundle A group of optical fibers randomly positioned but essentially parallel to each other in a bundle that is used simply as a means of guiding beams of light, with no particular spatial relationship among the fibers at the beginning or at the end, or between the ends. *Synonym: LIGHT CONDUIT.*

nonlinear distortion Distortion caused by a deviation from a linear relationship between specified input and output parameters of a system or component. *See also: LINEARITY.*

nonlinear scattering Direct conversion of a photon from one wavelength to one or more other wavelengths. In an optical waveguide, nonlinear scattering is usually not important below the threshold irradiance for stimulated nonlinear scattering.
Note: Examples are Raman and Brillouin scattering. *See also: PHOTON.*

nonradiative recombination In an electroluminescent diode in which electrons and holes are injected into the p-type and n-type regions by application of a forward bias, the recombination of injected minority carriers with the majority carriers in such a manner that the energy released upon recombination results in heat, which is dissipated primarily by conduction and some thermal radiation.
Note: Energy released by nonradiative recombination in LEDs does not contribute to light energy for optical use such as energizing optical fibers or driving integrated optical circuits. *See also: RADIATIVE RECOMBINATION.*

nonreflective coupler An optical fiber coupling device that enables signals in one or more fibers to be transmitted to one or more other fibers by entering the input signal fibers into an optical fiber volume without an internal reflecting surface so that the diffused signals pass directly to the output fibers on the opposite side of the fiber volume for conduction away in one or more of the output fibers.

Note: The optical fiber volume is a shaped piece of the optical fiber material to achieve transmission of two or more inputs to two or more outputs. *See also: REFLECTIVE STAR-COUPLER; TEE COUPLER.*

non-return-to-zero (NRZ) code A code form having two states termed zero and one, and no neutral or rest condition. *See also: RETURN-TO-ZERO CODE.*

nonsynchronous data transmission channel A data channel in which no separate timing information is transferred between the DTE and the DCE. *See also: CLOCK; NETWORK.*

nonsynchronous network A network in which the clocks need be neither synchronous nor mesochronous. *Synonym: ASYNCHRONOUS NETWORK.*

normal emergence In optics, a condition in which a ray emerges along the normal to the emergent surface of a medium. *See also: EMERGENCE; GRAZING EMERGENCE.*

normal incidence Pertaining to light rays incident at 90-degrees to the incident surface.

normalized frequency A dimensionless quantity (denoted by V), given by:

$$V = \frac{2\pi a}{\lambda} \sqrt{n_1^2 - n_2^2}$$

where a is waveguide core radius, λ is wavelength in vacuum, and n_1 and n_2 are the maximum refractive index in the core and refractive index of the homogeneous cladding, respectively. In a fiber having a power-law profile, the approximate number of bound modes is $(V^2/2)[g/(g + 2)]$, where g is the profile parameter. *Synonym: V NUMBER. See also: BOUND MODE; MODE VOLUME; PARABOLIC PROFILE; POWER-LAW INDEX PROFILE; SINGLE MODE OPTICAL WAVEGUIDE.*

normalized waveguide thickness In planar symmetric dielectric waveguides

of the type used in diode lasers, a nondimensional quantity characterizing the waveguide. Symbol: D.

Note: $D \equiv (2\pi d/\lambda) \sqrt{n_1^2 - n_2^2}$, where d is the active-layer thickness, λ is the vacuum wavelength, n_1 and n_2 are the bulk refractive indices of the active and confinement layers, respectively.

Note: The equivalent of the normalized frequency used in characterizing fiber guides.

Note: For examples see D. Botez and G. Herskowitz, *Proc. IEEE,* vol. 68, pp. 689–732, June 1980.

normal magnification *See: INDIVIDUAL NORMAL MAGNIFICATION.*

notched noise Noise in which a narrow band of frequencies has been removed.

not-ready condition A steady-state condition at the DTE/DCE interface that denotes that the DCE is not ready to accept a call request signal or that the DTE is not ready to accept an incoming call, respectively.

Note: The not-ready condition may be controlled or uncontrolled.

NRZ *Acronym for NON-RETURN-TO-ZERO CODE.*

NRZI recording ("Non-return-to-zero, change-on-ones"). A method of magnetic recording in which ones are represented by a change in condition and zeros are represented by no change.

NS laser *See: NARROW-STRIPE LASER.*

n-type semiconductor material Semiconductor material for which the inclusion of dopant(s) creates excess electrons. The dopant atoms providing electrons are called donors. The substrates for the vast majority of optoelectronic devices are n-type.

number *See: T-NUMBER.*

numerical aperture (NA) 1. The sine of the vertex angle of the largest cone of meridional rays that can enter or leave an optical system or element, multiplied by the refractive index of the medium in which the vertex of the cone is located. Generally measured with respect to an object or image point and will vary as that point is moved. 2. For an optical fiber in which the refractive index decreases monotonically from n_1 on axis to n_2 in the cladding the numerical aperture is given by:

$$NA = \sqrt{n_1^2 - n_2^2}$$

3. Colloquially, the sine of the radiation

or acceptance angle of an optical fiber, multiplied by the refractive index of the material in contact with the exit or entrance face. This usage is approximate and imprecise, but is often encountered.

See also: ACCEPTANCE ANGLE; LAUNCH NUMERICAL APERTURE; MERIDIONAL RAY; RADIATION ANGLE; RADIATION PATTERN.

NU value *See: ABBE CONSTANT.*

object In an optical system, the figure viewed through or imaged by an optical system.

Note: It may consist of natural or artificial structures or targets, or may be the real or virtual image of an object formed by another optical system. In the optical field, an object should be thought of as an aggregation of points. *See: SINEWAVE OBJECT.*

objective In an optical system such as microscopes and telescopes, the optical component that receives light from the object and forms the first or primary image.

Note: In cameras, the image formed by the objective is the final image. In telescopes and microscopes, when used visually, the image formed by the objective is magnified by use of an eyepiece.

object plane In an optical system, the plane that contains the object points lying within the field of view.

occluder In optics, a device that completely or partially limits the amount of light reaching the eye.

OCR *Acronym for OPTICAL CHARACTER RECOGNITION.*

octet A group or byte of eight binary digits usually operated upon as an entity.

octet alignment Alignment of bits into sequences of 8 binary digits.

octet timing signal A signal that identifies each octet in a contiguous sequence of serially transmitted octets.

odd-even check *Synonym: PARITY CHECK.*

off-axis paraboloidal mirror A paraboloidal mirror that consists of only a portion of a paraboloidal surface through which the axis does not pass.

off-net calling That process wherein telephone calls which originate or pass through private switching systems in transmission networks are extended to stations in the commercial telephone system.

OH-content Quantity expressing the amount in ppm or ppb of hydroxyl radical in silica-based fibers. The lower the OH-content, the closer the fiber attenuation loss spectrum is to the ultimate loss for silica-based fibers.

Note: See for example E. Iwahashi, *IEEE J. Quantum Electron.,* vol. QE-17, pp. 890–897, June 1981.

OM-CVD *See: OM-VPE.*

OM-VPE *Acronym for ORGANO-METALLIC VAPOR-PHASE EPITAXY.*

one-way communication A mode of communication such that information is always transferred in one preassigned direction only.

one-way-only channel A channel capable of operation in only one direction, which is fixed and cannot be reversed. *Synonym: UNIDIRECTIONAL CHANNEL.*

one-way reversible operation *Synonym: HALF-DUPLEX OPERATION.*

one-way trunk A trunk between switching centers used for traffic in one preassigned direction.

Note: One-way trunks are normally used to collect a particular switching center's originating traffic for transmission to a particular destination. At the originating end the one-way trunk is known as an outgoing trunk; at the other end, it is known as an incoming trunk.

opaque 1. Impervious to light, i.e., has zero luminous transmittance. 2. A substance that is impervious to light applied to transparent or translucent substances. 3. To make impervious to light.

open waveguide A waveguide in which electromagnetic waves are guided by a gradient in the refractive index such that the waves are confined to the guide by refraction within, or reflection from, the outer surface of the guide, thus the electromagnetic waves propagate, without radiation, along the interface between different media; for example, an optical fiber. *See also: CLOSED WAVEGUIDE.*

optic *See: ACOUSTO-OPTIC; ELECTRO-OPTIC.*

optical adaptive technique *See: CO-HERENT OPTICAL ADAPTIVE TECH-NIQUE.*

optical attenuator In an optical fiber data link or integrated optical circuit, a device used to reduce the intensity, i.e. attenuate the lightwaves when inserted into an optical waveguide.

Note: Three basic forms of optical attenuators have been developed; namely a fixed optical attenuator, a stepwise variable optical attenuator, and a continuous variable attenuator. One form of attenuator uses a filter consisting of a metal film evaporated onto a sheet of glass to obtain the attenuation. The filter might be tilted to avoid reflection back into the input optical fiber or cable. *See also: CONTINUOUS VARIABLE OPTICAL ATTENUATOR; FIXED OPTICAL AT-TENUATOR; STEPWISE VARIABLE OP-TICAL ATTENUATOR.*

optical axis In an optical waveguide, synonymous with "fiber axis."

optical bandwidth For light-emitting diodes, the frequency at which the optical output drops by 3 dB with respect to its low-frequency value.

optical blank A casting consisting of an optical material molded into the desired geometry for grinding, polishing, or (in the case of optical waveguides) drawing to the final optical/mechanical specifications. *See also: PREFORM.*

optical cable A fiber, multiple fibers, or fiber bundle in a structure fabricated to meet optical, mechanical, and environmental specifications. *Synonym: OPTI-CAL FIBER CABLE. See also: FIBER BUN-DLE; OPTICAL CABLE ASSEMBLY.*

optical cable assembly An optical cable that is connector terminated. Generally, an optical cable that has been terminated by a manufacturer and is ready for installation. *See also: FIBER BUNDLE; OPTICAL CABLE.*

optical cavity A region bounded by two or more reflecting surfaces, referred to as mirrors, end mirrors, or cavity mirrors, whose elements are aligned to provide multiple reflections. The resonator in a laser is an optical cavity. *Synonym: RESO-NANT CAVITY. See also: ACTIVE LASER MEDIUM; LASER.*

optical-cavity diode *See: LARGE OPTI-CAL-CAVITY DIODE.*

optical cement A permanent and transparent adhesive capable of withstanding extremes of temperature.

Note: Canada Balsam is a classic optical cement although it is being replaced by modern synthetic adhesives, such as methacrylates, caprinates, and epoxies.

optical character recognition (OCR) The machine identification of printed characters through use of light-sensitive devices.

optical circuit *See: INTEGRATED OPTI-CAL CIRCUIT.*

optical circuit filter-coupler-switch modulator *See: INTEGRATED-OPTI-CAL CIRCUIT FILTER-COUPLER-SWITCH MODULATOR.*

optical combiner A passive device in which power from several input fibers is distributed among a smaller number (one or more) of input fibers. *See also: STAR COUPLER.*

optical conductor *Deprecated synonym for OPTICAL WAVEGUIDE.*

optical conductor loss *See: CONNEC-TOR-INDUCED OPTICAL-CONDUC-TOR LOSS.*

optical connector *See: OPTICAL WAVEGUIDE CONNECTOR.*

optical contact A condition in which two sufficiently clean and close fitting surfaces adhere together without reflection at the interface.

Note: The optically contacted surface is practically as strong as the body of the glass.

optical coupler *See: OPTICAL WAVE-GUIDE COUPLER.*

optical data bus An optical fiber network, interconnecting terminals, in which any terminal can communicate with any other terminal. *See also: OPTI-CAL LINK.*

optical data link A system consisting of a transmitter, i.e., a light source; a fiber optic cable; and a receiver, i.e., a photodetector, all connected together in such a manner that light waves from the source can be received at the receiver.

Note: Light from the transmitter is usually modulated by an intelligence-bearing signal.

optical density A measure of the transmittance of an optical element expressed

by: $\log_{10}(1/T)$ or $-\log_{10}T$, where T is transmittance. The analogous term $\log_{10}(1/R)$ is called reflection density.
Note: The higher the optical density, the lower the transmittance. Optical density times 10 is equal to transmission loss expressed in decibels; for example, an optical density of 0.3 corresponds to a transmission loss of 3 dB. *See also: TRANSMISSION LOSS; TRANSMITTANCE.*

optical detector A transducer that generates an output signal when irradiated with optical power. *See also: OPTOELECTRONIC.*

optical directional coupler A device used in optical fiber communication systems, such as CATV and data-links for optical fiber measurements, to combine or split optical signals at desired ratios by insertion into a transmission line. For example, a three-port or four-port unit with precise connectors at each port to enable inputs to be coupled together and transmitted via multiple outputs.

optical dispersion attenuation The attenuation of a signal in an optical waveguide, caused by the fact that each frequency component of a launched pulse is attenuated such that higher frequencies are attenuated more than the lower frequencies, giving rise to attenuation distortion.
Note: The dispersion attenuation factor is given as

$$e^{-AF^2}$$

where A is a material constant, including substance and geometry, and F is a frequency component of the signal being attenuated.

optical distortion An aberration of spherical surface optical systems due to the variation in magnification with distance from the optical axis.

optical emitter A source of optical power, that is, a source of electromagnetic radiation in the visible and near-visible region of the frequency spectrum.

optical end-finish The surface condition at the face of an optical conducting medium.

optical energy density The energy in a light beam passing through a unit area normal to the direction of propagation or the direction of maximum power gradient, expressed in joules per square meter.

optical fiber Any filament or fiber, made of dielectric materials, that guides light, whether or not it is used to transmit signals. *See also: FIBER BUNDLE; FIBER OPTICS; OPTICAL WAVEGUIDE.*

optical fiber bundle Many optical fibers in a single protective sheath or jacket.
Note: The jacket is usually polyvinyl chloride (PVC). The number of fibers might range from a few to several hundred, depending on the application and the characteristics of the fibers.

optical fiber cable *Synonym: OPTICAL CABLE.*

optical fiber coating A protective material bonded to an optical fiber, over the cladding if any, to preserve fiber strength and inhibit cabling losses, by providing protection against mechanical damage, protection against moisture and debilitating environments, compatibility with fiber and cable manufacture, and compatibility with the jacketing process.
Note: Coatings include fluorpolymers, Teflon, Kynar, polyurethane, and many others. Application methods include dipcoating (for those in solution), extrusion, spray coating, and electrostatic coating.

optical fiber jacket The material that covers the buffered or unbuffered optical fiber.

optical fiber junction An interface formed by joining the ends of two optical fibers in a coaxial (in-line) butt joint for direct fiber-to-fiber transmission.

optical fiber preform Optical fiber material, such as silica, optical glass, or plastic, in a particular shape, such as a rod or hollow tube, from which an optical fiber is made usually by drawing or rolling, for example a solid glass rod made with a higher refractive-index than the tube into which it is slipped, to be heated and drawn or rolled into a cladded optical fiber; or four lower-refractive-index rods surrounding a higher-refractive-index rod heated and drawn into a cladded fiber.
Note: The drawing process results in fiber many times longer than the preforms.

optical fiber transfer function The transformation that an optical fiber brings about on an electromagnetic wave that enters it, such that if the input signal composition is known, and the transfer function is known for the fiber, the output signal can be determined: for example:

$$FTF = e^{-AF^2}$$

where FTF is the fiber transfer function, A is a constant for the fiber, and F is a frequency component of the signal, i.e., of the wave.

optical-fiber trap A hair-fine optical fiber that is nearly invisible that breaks easily when strained that can be placed on fences or in fields, that can signal the location of a break and thus cannot be cut without detection, and thus can be used to warn of trespassers. *Synonym: SECURITY OPTICAL FIBER.*

optical fiber waveguide *Synonym: OPTICAL WAVEGUIDE.*

optical filter An element that selectively transmits or blocks a range of wavelengths.

optical frequency division multiplex *See: WAVELENGTH DIVISION MULTIPLEX (WDM).*

optical harness A number of multiple fiber cables or jacketed bundles placed together in an array that contains branches.
Note: A harness is usually installed within other equipment and mechanically secured to that equipment.

optical harness assembly An optical harness that is terminated and ready for installation.

optical integrated circuit *See: INTEGRATED OPTICAL CIRCUIT.*

optical isolator In fiber-optical systems a device interposed between the light source and the fiber, that prevents light reflected at the fiber receiving end or from anywhere down the line to reach the diode laser. Source isolation prevents excess noise due to reflections.
Note: See for example K. Kobayashi *et al., IEEE J. Quantum Electron.,* vol. QE-16, pp. 11–23, Jan. 1980.

optical lever The means of amplifying small angular movements by reflecting a beam of light from a mirror or prism.

optical link Any optical transmission channel designed to connect two end terminals or to be connected in series with other channels.
Note: Sometimes terminal hardware (e.g., transmitter/receiver modules) is included in the definition. *See also: OPTICAL DATA BUS.*

optically active material A material

that can rotate the polarization of light that passes through it.
Note: An optically active material exhibits different refractive indices for left and right circular polarizations (circular birefringence). *See also: BIREFRINGENT MEDIUM.*

optical maser A source of nearly monochromatic and coherent radiation produced by the synchronous and cooperative emission of optically pumped ions introduced into a crystal host lattice, gas, or liquid atoms excited in a discharge tube, the radiation being a sharply defined frequency that propagates in an intense highly directional beam.

optical modulation *See: EXTERNAL OPTICAL MODULATION.*

optical modulator *See: THIN-FILM OPTICAL MODULATOR.*

optical multimode dispersion A dispersion, i.e., frequency distortion, of pulses in an optical waveguide caused by mode mixing when two or more transmission modes are supported by the same fiber.
Note: Optical multimode dispersion is greatly reduced in graded-index fibers, and somewhat reduced by using a monochromatic light source, such as a laser.

optical multiplexers *See: THIN-FILM OPTICAL MULTIPLEXERS.*

optical parametric oscillator A tunable device, usually a crystal, that varies the wavelength of a light beam from a solid state laser.

optical path length In a medium of constant refractive index n, the product of the geometrical distance and the refractive index. If n is a function of position,

$$\text{optical path length} = \int n \, ds$$

where ds is an element of length along the path.
Note: Optical path length is proportional to the phase shift a light wave undergoes along a path. *See also: OPTICAL THICKNESS.*

optical power *Colloquial synonym for RADIANT POWER.*

optical power density The energy per unit time transmitted by a light beam through a unit area normal to the direction of propagation or the direction of maximum power gradient, expressed in watts per square meter or joules per second (square-meter).

optical power efficiency The ratio of emitted electromagnetic power of an optical source to the electrical input power to the source.

optical protective coating Films that are applied to a coated or uncoated optical surface primarily for protecting the surface from mechanical abrasion, chemical corrosion, or both.

Note: An important class of protective coatings consists of evaporated thin films of titanium dioxide, silicon monoxide or magnesium fluoride. A thin layer of silicon monoxide may be added to protect an aluminized surface to prevent corrosion.

optical receiver A detector, capable of demodulating a light wave, that can be coupled to a transmission medium such as an optical fiber.

optical repeater In an optical waveguide communication system, an optoelectronic device or module that receives a signal, amplifes it (or, in the case of a digital signal, reshapes, retimes, or otherwise reconstructs it) and retransmits it. *See also: MODULATION.*

optical rotation 1. The angular displacement of the plane of polarization of light passing through a medium. 2. The azimuthal displacement of the field of view achieved through the use of a rotating prism.

optical space-division multiplexing (OSDM) The use of independent fibers contained in a bundle to provide optical paths for independent channels.

Note: Usually the fibers are collected into a single cable, with sheathing and strength members.

optical spectrum Generally, the electromagnetic spectrum within the wavelength region extending from the vacuum ultraviolet at 40 nm to the far infrared at 1 mm. *See also: INFRARED; LIGHT.*

optical surface In an optical system, a reflecting or refracting surface of an optical element, or any other identified geometric surface in the system.

Note: Normally, optical surfaces occur at surfaces of discontinuity (abrupt changes) in fractive indices, absorptive qualities, transmissivity, vitrification, or other optical quality or characteristic.

optical switch A switching circuit that enables signals in optical fibers and inte-

grated optical circuits (IOC) to be selectively switched from one circuit or path to another or to perform logic operations with these signals. By using electro-optic effects, magneto-optic effects, or other effects or methods that operate on transmission modes, such as, transverse-electric versus transverse magnetic modes, type and direction of polarization, or other characteristics of electromagnetic waves. *See also: THIN-FILM OPTICAL SWITCH; FLIP-CHIP.*

optical taper An optical fiber with a diameter that is a linear function of its length, thus having a conical shape, either increasing or decreasing in diameter in the direction of propagation of a light wave in a longitudinal direction.

Note: The taper can be used to increase or decrease the size of an image when the tapers are bundled. *Synonym:* CONICAL FIBER.

optical thickness The physical thickness of an isotropic optical element, times its refractive index. *See also: OPTICAL PATH LENGTH.*

optical time-domain reflectometry Technique for locating optical cable damage or faults. A high-pulsed-power laser is used to send pulses down the optical line. The backscattered light is detected and then processed with a box-car integrator. An analysis of the return waveform provides the location of faults and in-line connectors. *See: OTDR.*

optical transimpedance In an optical transmission system, the ratio of the output voltage at the detector end to the input current at the source end.

optical transmitter A source of light capable of being modulated and coupled to a transmission medium such as an optical fiber or an integrated optical circuit.

optical video disc (OVD) A disc on the surface of which digital data is recorded at high packing densities in concentric circles or in a spiral using a laser beam to record spots, that are read by means of a reflected laser beam of lower intensity than the recording intensity.

Note: Up to 10^{11} bits are being recorded on a single disc, thus being suitable for an hour of TV programming playback.

optical waveguide 1. Any structure capable of guiding optical power. 2. In optical communications, generally a fiber designed to transmit optical signals. *Synonyms: LIGHTGUIDE; OPTICAL CON-*

DUCTOR (DEPRECATED); OPTICAL FI-
BER WAVEGUIDE. See also: CLADDING;
CORE; FIBER BUNDLE; FIBER OPTICS;
MULTIMODE OPTICAL WAVEGUIDE;
OPTICAL FIBER; SINGLE MODE WAVE-
GUIDE; TAPERED FIBER WAVEGUIDE.

optical waveguide connector A device whose purpose is to transfer optical
power between two optical waveguides
or bundles, and that is designed to be
connected and disconnected repeatedly.
See also: MULTIFIBER JOINT; OPTICAL
WAVEGUIDE COUPLER.

optical waveguide coupler 1. A device
whose purpose is to distribute optical
power among two or more ports. See also:
STAR COUPLER; TEE COUPLER. 2. A
device whose purpose is to couple optical power between a waveguide and a
source or detector.

optical waveguide preform See: PRE-
FORM.

optical waveguide splice A permanent
joint whose purpose is to couple optical
power between two waveguides.

optical waveguide termination A
configuration or a device mounted at the
end of a fiber or cable which is intended
to prevent reflection. See also: INDEX
MATCHING MATERIAL.

optic axis In an anisotropic medium, a
direction of propagation in which orthogonal polarizations have the same
phase velocity. Distinguished from "optical axis." See also: ANISOTROPIC.

optic bundle See: FIBER-OPTIC BUN-
DLE.

optic cable See: FIBER-OPTIC CABLE.

optic communications See: FIBER-OP-
TIC COMMUNICATIONS.

optic connector See: FIXED FIBER-OPTIC
CONNECTOR; FREE FIBER-OPTIC CON-
NECTOR.

optic-electronic device See: OPTO-
ELECTRONIC DEVICE.

optic multiport coupler See: FIBER-OP-
TIC MULTIPORT COUPLER.

optic probe See: FIBER-OPTIC PROBE.

optic rod coupler See: FIBER-OPTIC
ROD COUPLER.

optic rod multiplexer-filter See: FI-
BER-OPTIC ROD MULTIPLEXER-FIL-
TER.

optics That branch of physical science
concerned with the nature and properties of electromagnetic radiation and
with the phenomena of vision. See:
COATED OPTICS; ELECTRON OPTICS;
FIBER OPTICS; GEOMETRIC OPTICS;

PHYSICAL OPTICS; INTEGRATED OP-
TICS. See: ACTIVE OPTICS; CRYSTAL
OPTICS; GEOMETRIC OPTICS; ULTRA-
VIOLET FIBER OPTICS.

optic scrambler See: FIBER-OPTIC
SCRAMBLER.

optic splice See: FIBER-OPTIC SPLICE.

optic terminus See: FIBER OPTIC TER-
MINUS.

optic transmission system See: FIBER-
OPTIC TRANSMISSION SYSTEM; LA-
SER FIBER OPTIC TRANSMISSION SYS-
TEM.

optic waveguide See: FIBER-OPTIC
WAVEGUIDE.

optoelectronic Pertaining to a device
that responds to optical power, emits or
modifies optical radiation, or utilizes optical radiation for its internal operation.
Any device that functions as an electri-
cal-to-optical or optical-to-electrical
transducer.
Note: 1. Photodiodes, LEDs, injection lasers and integrated optical elements are
examples of optoelectronic devices commonly used in optical waveguide communications.
Note: 2. "Electro-optical" is often erroneously used as a synonym. See also:
ELECTRO-OPTIC EFFECT; OPTICAL DE-
TECTOR.

order See: DIFFRACTION GRATING
SPECTRAL ORDER.

order of diversity The number of independently fading propagation paths or
frequencies, or both, used in diversity
reception. See also: DIVERSITY RECEP-
TION; DUAL DIVERSITY; QUADRUPLE
DIVERSITY.

ordinary ray The ray that has an isotropic velocity in a doubly refracting crystal, obeying Snell's Law upon refraction
at the crystal surface.

organo-metallic vapor phase epitaxy
Single-crystal growth technique during
which atomic and molecular species obtained from organic compounds are carried in the gas phase and deposited onto
a single-crystal substrate. The deposited
material is a product of chemical reactions between the various gaseous compounds. Symbols: OM-VPE, OM-CVD,
MO-CVD. OM-VPE provides crystalline layers very uniform in thickness.
Promising technique for high-volume
fabrication of LEDs, detectors and low-
power lasers.

orthogonal multiplex A method of
TDM in which pulses with orthogonal

properties are used so as to avoid inter-symbol interference.

oscillator *See: OPTICAL PARAMETRIC OSCILLATOR.*

OSDM *Acronym for OPTICAL SPACE-DIVISION MULTIPLEXING.*

OTDR *Acronym for OPTICAL TIME-DOMAIN REFLECTOMETRY.*

out-of-band signaling Signaling which utilizes frequencies within the guard band between channels, or utilizes bits other than information bits in a digital system. This term is also used to indicate the use of a portion of the channel bandwidth provided by the medium such as the carrier channel, but denied to the speech or intelligence path by filters. It results in a reduction of the effective available bandwidth.

output angle *Synonym: RADIATION ANGLE.*

output rating 1. The power available at the output terminals of a transmitter when connected to the normal load or its equivalent. 2. Under specified ambient conditions, the power that can be delivered by a device over a long period of time without overheating.

outside vapor phase oxidation process (OVPO) A CVPO process, for the production of optical fiber, in which the soot stream, and heating flame, is deposited on the outside surface of the rotating glass rod. *See also: INSIDE VAPOR PHASE OXIDATION PROCESS; CHEMICAL VAPOR PHASE OXIDATION PROCESS.*

OVD *Acronym for OPTICAL VIDEO DISC.*

overmodulation A condition in which the mean level of the modulating signal is such that the peak value of the signal exceeds the value necessary to produce 100% modulation, resulting in distortion of the output signal.

overshoot 1. The result of an unusual atmospheric condition that causes microwave signals to be received where they are not intended. 2. In an amplifier, the increased amplitude of a portion of a non-sinusoidal wave due to the particular characteristics of the circuit.
Note: Overshoot is valuable in decreasing the response time of a signal, but it causes distortion of that signal.

OVPO *Acronym for OUTSIDE VAPORPHASE OXIDATION PROCESS.*

oxidation process *See: AXIAL VAPORPHASE OXIDATION PROCESS; CHEMICAL VAPOR-PHASE OXIDATION PROCESS; INSIDE VAPOR-PHASE OXIDATION PROCESS; MODIFIED INSIDE VAPOR-PHASE OXIDATION PROCESS; OUTSIDE VAPOR-PHASE OXIDATION PROCESS.*

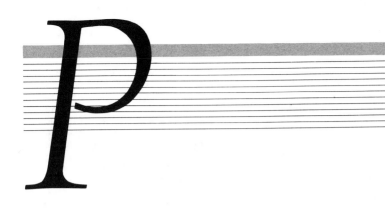

packet 1. A group of data and control characters in a specified format, transferred as a whole. 2. A group of binary digits, including data and call control signals, which is switched as a composite whole. The data, all control signals, and possibly error control information, are arranged in a specific format. *See also: ENVELOPE.*

packet format The structure of data and control information in a packet.
Note: The size and content of the various fields are defined by a set of rules that are used to make up a packet.

packet mode terminal Data terminal equipment that can control, format, transmit, and receive packets.

packet-switched data transmission service A service that provides the transmission of data in the form of packets.
Note: This service may or may not provide for the assembly and disassembly of data packets. *See also: DATA TRANSMISSION.*

packet switching 1. A system in which messages are broken down into smaller units called packets, which are then individually addressed and routed through the network. 2. The process of routing and transferring data by means of addressed packets so that a channel is occupied during the transmission of the packet only, and upon completion of the transmission the channel is made available for the transfer of other traffic. *See also: MESSAGE SWITCHING; SWITCHING SYSTEM.*

packing density In a bundle of fibers, the end cross-sectional total core area per unit of cross-sectional area of the assembly of fibers whose cross-sectional core areas are counted.
Note: Packing density varies with size of fiber, core areas relative to total fiber,

the geometric or spatial distribution of fibers, the overall size of fibers, tightness of packing, and other factors.

packing fraction loss The power loss, expressed in decibels (dB), due to the packing fraction.

packing fraction (PF) In a fiber bundle, the ratio of the aggregate fiber cross-sectional core area to the total cross-sectional area (usually within the ferrule) including cladding and interstitial areas. *See also: FERRULE; FIBER BUNDLE.*

PACVD *Acronym for PLASMA-ACTIVATED CHEMICAL VAPOR DEPOSITION PROCESS.*

PAM *PULSE-AMPLITUDE MODULATION.*

pancratic lens *See: ZOOM LENS.*

parabolic index profile A power-law index profile with the profile parameter, g, equal to 2. *Synonym: QUADRATIC PROFILE. See also: GRADED INDEX PROFILE; MULTIMODE OPTICAL WAVEGUIDE; POWER-LAW INDEX PROFILE; PROFILE PARAMETER.*

paraboloidal mirror A concave mirror that has the form of a paraboloid of revolution.
Note: The paraboloidal mirror may consist of only a portion of a paraboloidal surface through which the axis does not pass. It is known as an off-axis paraboloidal mirror. All axial parallel light rays are focused at the focal point of the paraboloid without spherical aberration, and conversely all light rays emanating from an axial source at the focal point are reflected as a bundle of parallel rays without spherical aberration. Paraboloidal mirrors are free from chromatic aberration. *See also: OFF-AXIS PARABOLOIDAL MIRROR.*

parallel light *See: COLLIMATED LIGHT.*

parallel-to-serial converter 1. A device that converts a group of parallel inputs, all of which are presented simultaneously, into a corresponding time sequence of signal elements. 2. A device that converts a spatial distribution of signal states representing data into a corresponding time sequence of signal states.

Note: Each of the parallel input signals requires a separate channel while the serial output requires only a single channel. *See also: SERIAL-TO-PARALLEL CONVERTER.*

parallel transmission The simultaneous transmission of a certain number of signal elements.

Note: For example, use of a code in which each signal element is characterized by a combination of 3 out of 12 frequencies simultaneously transmitted over the channel, or use of a separate wire or circuit for each signal element of a character, or word, so that the signal elements of a character or word are simultaneously transmitted. *See also: SERIAL TRANSMISSION.*

parametric oscillator *See: OPTICAL PARAMETRIC OSCILLATOR.*

paraxial ray A ray that is close to and nearly parallel with the optical axis.

Note: For purposes of computation, the angle, θ, between the ray and the optical axis is small enough for sin θ or tan θ to be replaced by θ (radians). *See also: LIGHT RAY.*

parity In binary-coded systems, a condition obtained with a self-checking code such that in any permissible code expression the total number of 1s or 0s is always even or always odd.

parity bit A check bit appended to an array of binary digits to make the sum of all the binary digits, including the check bit, always odd or always even. *See also: CHECK DIGIT.*

parity check A check that tests whether the number of 1s (or 0s) in an array of binary digits is odd or even.

Note: Odd parity is standard for synchronous transmission and even parity for asynchronous transmission. *Synonym: ODD-EVEN CHECK.*

parts per billion Quantity expressing how many grams of a given impurity correspond to one billion grams of fiber material. Used mostly for expressing the OH-content in silica-based fibers. Symbol: ppb.

parts per million Quantity expressing how many grams of a given impurity

correspond to one million grams of fiber material. Symbol: ppm.

pass band The number of hertz expressing the difference between the limiting frequencies at which the desired fraction (usually half power) of the maximum output is obtained. Term applies to all types of equipment and circuits.

passive repeater An unpowered device used to route a microwave beam over or around an obstruction.

Note: For example, two parabolic antennas connected back-to-back, or a flat reflector used as a mirror.

path clearance In microwave line-of-sight communication, the perpendicular distance from the radio beam axis to obstructions such as trees, buildings, or terrain.

Note: The required path clearance is usually expressed, for a particular K-factor, as some fraction of the first Fresnel zone radius. *See also: EFFECTIVE EARTH RADIUS; FRESNEL ZONE; K-FACTOR; PATH PROFILE; PATH SURVEY; PROPAGATION PATH OBSTRUCTION.*

path intermodulation noise Noise in a transmission path contingent upon modulation; it results from any nonlinear characteristic of the path.

path length *See: OPTICAL PATH LENGTH.*

path loss The decrease in power in transmission from one point to another.

Note: It is usually expressed in decibels, and in radio systems is taken to be the loss between transmitting and receiving antennas.

path profile A graphic representation of a propagation path, showing the surface features of the earth, such as trees, buildings, and other features that may cause obstruction or reflection, in the vertical plane containing the path. *See also: FRESNEL ZONE; K-FACTOR; PATH CLEARANCE; PROPAGATION PATH OBSTRUCTION; SMOOTH EARTH.*

path survey The assembling of pertinent geographical and environmental data required to design a microwave communication system. *See also: FRESNEL ZONE; K-FACTOR; PATH CLEARANCE.*

pattern *See: ACCEPTANCE PATTERN; RADIATION PATTERN.*

PCM *Acronym for PULSE-CODE MODULATION.*

PCM multiplex equipment Equipment for deriving a single digital signal at a

defined digit rate from two or more analog channels by a combination of pulse-
code modulation and time division multiplexing and also for carrying out the
inverse function, i.e., demultiplexing.
Note: The term should be preceded by
the relevant equivalent binary digit rate,
e.g., 2048 kbit/s PCM multiplex equipment. *See also: MULTIPLEX; PULSE
CODE MODULATION.*

PCS *Acronym for PLASTIC CLAD SILICA.*

PCW laser *See: PLANO-CONVEX
WAVEGUIDE DIODE LASER.*

PD *Acronym for PHOTODETECTOR.*

PDM *Acronym for PULSE-DURATION
MODULATION.*

peak *See: ABSORPTION PEAK.*

peak busy hour *See: BUSY HOUR; ER-
LANG.*

peak envelope power (PEP) The average power supplied to the antenna transmission line by a radio transmitter during one radio frequency cycle at the
highest crest of the modulation envelope, taken under conditions of normal
operation.

peak limiting A process in which the
absolute instantaneous value of a signal
parameter is prevented from exceeding a
specified value. *See also: CLIPPER; COM-
PANDOR; COMPRESSOR; EXPANDOR;
LIMITER; LIMITING.*

peak power output The output power
averaged over the radio frequency cycle
having the maximum peak value which
can occur under any combination of signals transmitted.

peak radiant intensity The maximum
value of radiant intensity of a lightwave.

peak-signal level An expression of the
maximum instantaneous signal power or
voltage as measured at any point in a
transmitted path. *See also: SIGNAL.*

peak-to-average ratio The ratio of the
instantaneous peak value (amplitude) of
a signal to its time averaged value.
Note: Peak-to-average ratio can be determined for voltage, current, power, or
other parameter.

peak-to-peak value The algebraic difference between the extreme values of a
varying quantity.

peak wavelength The wavelength at
which the radiant intensity of a source is
maximum. *See also: SPECTRAL LINE;
SPECTRAL WIDTH.*

pencil *See: LIGHT PENCIL.*

penta prism A prism having the unique
property of being able to divert a beam

of light ninety degrees in the principal
plane even if the beam does not strike
the end faces exactly normally.

PEP *Acronym for PEAK ENVELOPE POW-
ER.*

percentage modulation 1. In angle
modulation, the fraction of a specified
reference modulation, expressed in percent. 2. In amplitude modulation, the
modulation factor expressed in percent.
Note: It is sometimes convenient to ex-
press percentage modulation in dB below 100 percent modulation.

**peripheral strength-member optical
cable** A cable containing optical fibers
that are on the inside of a group of outer
high tensile-strength material such as
standard or solid contrahelical or longitudinal steel nylon, or other material,
with a crush-resistant jacketing (sheathing) on the outside of the cable. *See also:
CENTRAL STRENGTH-MEMBER OPTI-
CAL CABLE.*

permanent virtual circuit A virtual
circuit that is used to establish a long-
term association (between two DTEs) that
is identical to the data transfer
phase of a virtual call.
Note: Permanent virtual circuits eliminate
the need for repeated call setup and
clearing. *See also: VIRTUAL CALL CAPA-
BILITY; VIRTUAL CARRIER; VIRTUAL
CIRCUIT.*

PF *Acronym for PACKING FRACTION.*

PFM *Acronym for PULSE-FREQUENCY
MODULATION.*

phase *See: ACCESS PHASE; DATA
PHASE; DELAY EQUALIZER; DISEN-
GAGEMENT PHASE; INFORMATION
TRANSFER PHASE; NETWORK CON-
TROL PHASE; PHASED ARRAY; PHAS-
ING; PRE-DETECTION COMBINING.*

phase coherence *See: COHERENT.*

phase constant The imaginary part of
the axial propagation constant for a particular mode, usually expressed in radians per unit length. *See also: AXIAL
PROPAGATION CONSTANT.*

phase delay In the transfer of a single-
frequency wave from one point to another in a system, the time delay of the
part of the wave that identifies its phase.
Note: The phase delay may be expressed
by the ratio of the total phase shift in
seconds to the frequency in hertz, i.e.
seconds/hertz; in degrees; in radians; or
in wavelengths such as a decimal frac-
tion of a wavelength, e.g., $2 \times 10^{-6}\lambda$. *See
also: ABSOLUTE DELAY.*

phase deviation In phase (or angle) modulation, the peak difference between the instantaneous angle of the modulated wave and the angle of the carrier.
Note: In the case of a sinusoidal modulating function, the value of the phase deviation, expressed in radians, is equal to the modulation index. *See also: PHASE MODULATION.*

phase difference Time, in electrical degrees, by which one wave leads or lags another.

phase distortion *Synonym: DELAY DISTORTION.*

phase equalizer *See: DELAY EQUALIZER.*

phase-frequency distortion That form of distortion which occurs under either or both of the following conditions: (a) if the phase-frequency characteristic is not linear over the frequency range of interest; (b) if the zero-frequency intercept of the phase-frequency characteristic is not zero or an integral multiple of 2π radians.

phase hit In a transmission channel, a momentary disturbance caused by sudden phase changes in the signal.

phase interference fading The variation in signal amplitude produced by the interaction of two or more components with different relative phases. *See also: FADING.*

phase jitter A form of phase perturbation. *See also: JITTER.*

phase-lock loop An electronic servo system controlling an oscillator so that it maintains a constant phase angle relative to a reference signal source.

phase modulation (PM) A form of angle modulation in which the angle relative to the unmodulated carrier angle is varied in accordance with the instantaneous value of the amplitude of the modulating signal. *See also: FREQUENCY MODULATION; ELECTRO-OPTIC PHASE MODULATION; PHASE DEVIATION.*

phase oxidation process *See: CHEMICAL VAPOR-PHASE OXIDATION PROCESS; INSIDE VAPOR-PHASE OXIDATION PROCESS; MODIFIED INSIDE VAPOR-PHASE OXIDATION PROCESS.*

phase perturbation That phenomenon, from causes known or unknown, which results in a relative shifting (often quite rapid) in the phase of a signal. The shifting in phase may appear to be random, cyclic, or both.

Note: A similar phenomenon related to amplitude perturbation exists which is not sufficiently understood to be acceptably defined at this time. The amount of phase perturbation may be expressed in degrees with any cyclic component expressed in hertz. The instantaneous relative phase may or may not be significant. *See also: JITTER.*

phase shift The change in phase of a periodic signal with respect to a reference.

phase shift keying (PSK) A method of modulation used for digital transmission wherein the phase of the carrier is discretely varied in relation to a reference phase, or the phase of the previous signal element, in accordance with the data to be transmitted.

phase term In the propagation of an electromagnetic wave in a waveguide, such as an optical fiber or metal pipe, the term H in the expression for the exponential variation characteristic of guided waves

$$e^{-PZ} = e^{-Z(IH+A)}$$

that represents the phase change per unit of propagation distance of the wave, causing pulse distortion.
Note: The phase term H is dependent upon the permittivity and permeability of the material filling the guide, the frequency, and the modal characteristics of the propagating wave. *See also: ATTENUATION TERM; PROPAGATION CONSTANT.*

phase velocity For a particular mode, the ratio of the angular frequency to the phase constant. *See also: AXIAL PROPAGATION CONSTANT; COHERENCE TIME; GROUP VELOCITY.*

phosphorescence Luminescence of a material that occurs during, and for some time after, the material has been stimulated by radiation energy. *See also: LUMINESCENCE.*

photoconduction An increase in the electrical condition capability resulting from the absorption of electromagnetic radiation by the material.

photoconductive cell A device for detecting or measuring electromagnetic radiation intensity by variation of the conductivity of a substance caused by absorption of the radiation. *Synonym: PHOTORESISTIVE CELL, PHOTORESISTOR.*

photoconductive device A device that

makes use of photoconductivity, such as a photoconductive cell.

photoconductive effect The phenomenon in which some nonmetallic materials exhibit a marked increase in electrical conductivity upon absorption of photon energy. Photoconductive materials include gases (ionization) as well as crystals. They are used in conjunction with semiconductor materials that are ordinarily poor conductors but become distinctly conducting when subjected to photon absorption. The photons excite electrons into the conduction band where they move freely, resulting in good electrical conductivity. The conductivity increase is due to the additional free carriers generated when photon energies are absorbed in energy transitions. The rate at which free carriers are generated and the length of time they persist in conducting states (their lifetime) determines the amount of conductivity change.

photoconductive film A film of material whose electrical current-carrying ability is enhanced when illuminated by electromagnetic radiation, particularly in the visible region of the frequency spectrum.

photoconductive gain factor The ratio of the number of electrons per second flowing through a circuit containing a cube of semiconducting material, whose sides are of unit length, to the number of photons per second of incident electromagnetic radiation absorbed in this volume.

photoconductive meter An exposure meter in which a battery supplies power through a photoconductive cell to an electrical current measuring device, such as a milliammeter, to measure the intensity of radiation, such as light intensity, incident upon its active surface.

photoconductive photodetector A photodetector that makes use of the phenomenon of photoconductivity in its operation, thus it detects the presence of electromagnetic radiation particularly in the visible region of the frequency spectrum, by changing its electrical resistance in accordance with the intensity of the incident radiation, thus controlling the current flow from an applied bias voltage power source.
Note: Usually a source of voltage is needed to drive a current that will vary according to the variation in conductivity

resulting from the variation in incident electromagnetic radiation.

photoconductivity The conductivity increase exhibited by some nonmetallic materials, resulting from the free carriers generated when photon energy is absorbed in electronic transitions. The rate at which free carriers are generated, the mobility of the carriers, and the length of time they persist in conducting states (their lifetime) are some of the factors that determine the amount of conductivity change. *See also:* PHOTOELECTRIC EFFECT.

photoconductor A material, usually a nonmetallic solid, whose conductivity increases when it is exposed to electromagnetic radiation.

photocurrent The current that flows through a photosensitive device (such as a photodiode) as the result of exposure to radiant power. Internal gain, such as that in an avalanche photodiode, may enhance or increase the current flow but is a distinct mechanism. *See also:* DARK CURRENT; PHOTODIODE.

photodetector (PD) A device capable of extracting the information from an optical carrier, i.e., a thermal detector or a photon detector, the latter being used for communications more than the former. *See also:* PHOTOCONDUCTIVE PHOTODETECTOR; PHOTOELECTRO-MAGNETIC PHOTODETECTOR; PHOTOEMISSIVE PHOTODETECTOR; PHOTOVOLTAIC PHOTODETECTOR.

photodetector responsivity The ratio of the rms value of the output current or voltage of a photodetector to the rms value of the incident optical power input.
Note: In most cases, detectors are linear in the sense that the responsivity is independent of the intensity of the incident radiation. Thus, the detector response in amps or volts is proportional to incident optical power (watts). Differential responsivity applies to small variations in optical power. Optical detectors are square law detectors that respond to optical intensity, i.e., the square of the electromagnetic field associated with the optical radiation. They are linear in the sense that the response in volts or amps varies linearly with optical power input.

photodiode A diode designed to produce photocurrent by absorbing light. Photodiodes are used for the detection of optical power and for the conversion of optical power to electrical power. *See also: AVALANCHE PHOTODIODE (APD); PHOTOCURRENT; PIN PHOTODIODE.*

photodiode coupler   *See: AVALANCHE PHOTODIODE COUPLER; POSTIVE-IN-TRINSIC-NEGATIVE PHOTODIODE COUPLER.*

photodiode leakage current   Excess current in the reverse-bias current-voltage characteristic of photodiodes due to current leakage over the surface of the device (i.e., surface leakage) and current leakage in the bulk. Surface leakage can be minimized by passivation techniques. Bulk leakage is due to diffusion of minority carriers from the undepleted material (e.g., in Ge photodiodes) and to thermal generation and recombination of carrier pairs in the depletion region (e.g., in Si and III-V photodiodes). In addition, long-wavelength III-V photodiodes display excessive leakage currents at high voltages as a result of junction breakdown by tunneling. *Synonym: DARK CURRENT.*

photoeffect   *See: EXTERNAL PHOTOEFFECT; INTERNAL PHOTOEFFECT; EXTRINSIC INTERNAL PHOTOEFFECT; INTRINSIC INTERNAL PHOTOEFFECT.*

photoeffect detector   *See: EXTERNAL PHOTOEFFECT DETECTOR; INTERNAL PHOTOEFFECT DETECTOR.*

photoelectric effect   1. External photoelectric effect: The emission of electrons from the irradiated surface of a material. *Synonym: PHOTOEMISSIVE EFFECT.* 2. Internal photoelectric effect: Photoconductivity.

photoelectromagnetic effect   The production of a potential difference by virtue of the interaction of a magnetic field with a photoconductive material subjected to incident radiation.
Note: The incident radiation creates hole-electron pairs that diffuse into the material. The magnetic field causes the pair components to separate, resulting in a potential difference across the material. In most applications, the light is made to fall on a flat surface of an intermetallic semiconductor located in a magnetic field that is parallel to the surface, excess hole-electron pairs are created, and these carriers diffuse in the direction of the light but are deflected by the magnetic field to give a current flow through the semiconductor that is at right angles to both the light rays and the magnetic field. This is due to transverse forces acting on electrons and holes diffusing into the semiconductor from the surface. *Synonym: PHOTOMAGNETOELECTRIC EFFECT.*

photoelectromagnetic   photodetector   A photodetector that makes use of the photoelectromagnetic effect, namely uses an applied magnetic field.

photoemissive cell   A device that detects or measures radiant energy by measurement of the resulting emission of electrons from surface that has or displays a photoemissive effect.

photoemissive effect   *Synonym: (EXTERNAL) PHOTOELECTRIC EFFECT.*

photoemissive   photodetector   tube photometer   A photometer that uses a tube made of a photoemissive material.
Note: It is highly accurate, but requires electronic amplification, and is used mainly in laboratories.

photoemissivity   The property of a substance that causes it to emit electrons when electromagnetic radiation in the visible region of the frequency spectrum is incident upon it.
Note: Normally an electric field is applied to collect the emitted electrons.

photomagnetoelectric effect   *See: PHOTOELECTROMAGNETIC EFFECT.*

photometer   *See: PHOTOEMISSIVE TUBE PHOTOMETER.*

photometry   The science devoted to the measurement of the effects of electromagnetic radiation on the eye.
Note: Photometry is an outgrowth of psychophysical aspects, and involves the determination of visual effectiveness by considering radiated power and the sensitivity of the eye to the frequency in question.

photomultiplier   An electron tube that multiples the effect of incident electromagnetic radiation by accelerating emitted electrons and using them to impinge upon other surfaces, knocking out additional electrons, until a large electric current is produced for low incident radiation levels.
Note: The electron tube contains a photocathode, one or more dynodes, and an output electrode. Electrons emitted from the cathode are amplified by secondary emission from the dynodes. The original electron emission is thus cascaded by the secondary electrodes.

photon   A quantum of electromagnetic energy. The energy of a photon is hν where h is Planck's constant and ν is the optical frequency. *See also: NONLINEAR SCATTERING; PLANCK'S CONSTANT.*

photon detector A device that responds to incident photons, i.e., a device capable of signaling, with some reasonable probability of being correct, the absorption of a photon (quantum of light energy).
Note: The photon detector exhibits a change in property when it absorbs a photo, i.e., it is photoemissive, photoconductive, photovoltaic, or photoelectromagnetic.

photon lifetime Average time that the photon spends in the cavity prior to its loss by either intracavity absorption or transmission (i.e., emission) outside the cavity through the facets. In a Fabry-Perot cavity the photon lifetime is the reciprocal of the product between the cavity losses (internal + mirror losses) and the speed of light in the lasing medium. Typical values for diode lasers are 1–5 × 10^{-12} sec; which set the upper limit to the modulation capability of the devices. *Synonym: DECAY LIFETIME.*
Note: For example see H. Kressel and J. K. Butler, *Semiconductor Lasers and Heterojunction LEDs*, Academic Press, New York, 1977.

photon noise *Synonym: QUANTUM NOISE.*

photoresistive cell *See: PHOTOCONDUCTIVE CELL.*

photoresistor *See: PHOTOCONDUCTIVE CELL.*

phototronic photocell *See: PHOTOVOLTAIC PHOTOCELL.*

photovoltaic Pertaining to the capability of generating a voltage as a result of exposure to visible or other radiation.

photovoltaic cell A device that detects or measures radiant energy by the production of a source of voltage proportional to the incident radiation intensity.
Note: It is possible to operate a photovoltaic cell without an additional source of voltage, since it develops a voltage. The cell detects or measures electromagnetic radiation by generating a potential at a junction (barrier layer) between two types of material, upon absorption of radiant energy. *Synonyms: BARRIER-LAYER CELL; BARRIER-LAYER PHOTOCELL; BOUNDARY-LAYER PHOTOCELL; PHOTOTRONIC PHOTOCELL.*

photovoltaic effect The production of a voltage difference across a pn junction resulting from the absorption of photon energy. The voltage difference is caused by the internal drift of holes and electrons. *See also: PHOTON.*

photovoltaic meter An exposure cell in which a photovoltaic cell produces a current proportional to the light intensity, or area exposed, falling on the cell. This current is measured by a sensitive current-measuring device, such as a microammeter.

photovoltaic photodetector A photodetector that makes use of the photovoltaic effect.
Note: Usually a source of voltage is not needed for the photovoltaic photodetector, since it is its own source of voltage.

physical optics The branch of optics that treats light propagation as a wave phenomenon rather than a ray phenomenon, as in geometric optics.

pick-off coupling *See: TANGENTIAL COUPLING.*

piece-wise linear encoding *Synonym: SEGMENTED ENCODING LAW.*

pigtail A short length of optical fiber, permanently fixed to a component, used to couple power between it and the transmission fiber. *See also: LAUNCHING FIBER.*

pilot A signal, usually a single frequency, transmitted over a system for supervisory, control, synchronization, or reference purposes.
Note: Sometimes it is necessary to employ several independent pilot frequencies. Most radio relay systems use radio or continuity pilots of their own but transmit also the pilot frequencies belonging to the carrier frequency multiplex system. *See also: SIGNAL.*

PIN diode A junction diode doped in the forward direction positive, intrinsic, and negative, in that order.
Note: PIN diodes are used as photodetectors in fiber and integrated optical circuits.

PIN photodiode A diode with a large intrinsic region sandwiched between p- and n-doped semiconducting regions. Photons absorbed in this region create electron-hole pairs that are then separated by an electric field, thus generating an electric current in a load circuit.

Planck's constant The number h that relates the energy E of a photon with the frequency v of the associated wave through the relation $E = hv$. $h = 6.626 \times 10^{-34}$ joule-sec. *See also: PHOTON.*

Planck's law The quantum of energy (E)

associated with an electromagnetic field of frequency ν is

$$E = h\nu$$

where h is Planck's constant

$$(h = 6.626 \times 10^{-34} \text{ Joule-sec})$$

and E is the photon energy.

Note: The product of energy times the time is sometimes referred to as the action. Hence, h is sometimes referred to as the elementary quantum of action. Planck's law is the fundamental law of quantum theory and has direct application in optical communications (lightwave communications). It describes the essential concept of the quanta of electromagnetic energy.

plane *See: FOCAL PLANE; IMAGE PLANE; MERIDIAN PLANE; OBJECT PLANE.*

plane wave A wave whose surfaces of constant phase are infinite parallel planes normal to the direction of propagation.

planoconcave lens A lens with one surface plane, the other concave.

planoconvex waveguide (PCW) diode laser Mode-stabilized LOC-type diode laser structure grown by one-step liquid-phase epitaxy over a channel etched into the substrate. The guide layer assumes a planoconvex-lens-like shape in the lateral direction, while the active layer is of constant thickness. Typical reliable output power levels: 3–7 mW/facet cw.

Note: For example see T. Furuse, *et al., Proceedings of the 5th European Conf. on Optical Communications,* Paper 2.2, Amsterdam, Sept. 1979.

plano lens A lens having no curved surface, or whose two curved surfaces neutralize each other, so that it possesses no net refracting power.

plasma-activated chemical vapor deposition process (PACVD) A chemical vapor deposition (CVD) process for making graded-index (GI) optical fibers by depositing a series of thin layers of materials of different refractive indices on the inner wall of a glass tube as chemical vapors flow through the tube, using a microwave cavity to stimulate the formation of oxides by means of a nonisothermal plasma generated by the microwave resonant cavity.

plastic-clad silica fiber An optical waveguide having silica core and plastic cladding.

plesiochronous In TDM, the relationship between two signals such that their corresponding significant instants occur at nominally the same rate, any variations being constrained within a specified limit. *See also: ANISOCHRONOUS; ASYNCHRONOUS TRANSMISSION; HETEROCHRONOUS; HOMOCHRONOUS; ISOCHRONOUS; ISOCHRONOUS MODULATION; MESOCHRONOUS.*

PM *Acronym for PHASE MODULATION.*

p-n junction In a semiconductor device, the interface between p-type material and n-type material. When a positive voltage is applied to the p-region the junction is said to be forward-biased. When a positive voltage is applied to the n-side the junction is said to be reverse-biased. In optoelectronic devices, p-n junctions are formed by crystalline growth or by dopant diffusion.

Note: For examples see S. M. Sze, *Physics of Semiconductor Devices,* 2nd edition, John Wiley & Sons, New York, 1981.

Pockels cell A material, usually a crystal whose refractive index change is linearly proportional to an applied electric field, the material being configured so as to be part of another system, such as an optical path, the cell thus providing a means of modulating the light in the optical path.

point *See: FOCAL POINT; PRINCIPAL FOCUS POINT.*

points *See: EMISSION-BEAM-ANGLE-BETWEEN-HALF-POWER-POINTS.*

polariscope A combination of a polarizer and an analyzer used to detect birefringence in materials placed between them or to detect rotation in the plane of polarization caused by materials placed between them.

polarization That property of a radiated electromagnetic wave describing the time-varying direction and amplitude of the electric field vector; specifically, the figure traced as a function of time by the extremity of the vector at a fixed location in space, as observed along the direction of propagation.

Note: In general, the figure is elliptical and it is traced in a clockwise or counterclockwise sense. The commonly referenced circular and linear polarizations are obtained when the ellipse becomes a circle or a straight line, respectively.

Clockwise sense rotation of the electric vector is designated right-hand polarization, and counter-clockwise sense rotation is designated left-hand polarization.

polarization diversity Any method of diversity transmission and reception wherein the same information signal is transmitted and received simultaneously on orthogonally polarized waves with fade-independent propagation characteristics. *See also: DIVERSITY RECEPTION.*

polarization modulation The modulation of an electromagnetic wave in such a manner that the polarization of the carrier wave; such as the direction of polarization of the electric and magnetic fields, or their relative phasing, to produce changes in polarization angle in linear, circular, or elliptical polarization; is varied according to a characteristic of an intelligence-bearing input signal, such as a pulse-or-no-pulse digital signal. *Note:* In optical fibers or other wave guides, polarization shifts that are made in accordance with an input signal variation are a practical means of modulation.

polarized light A light beam whose electric vector vibrates in a direction that does not change, unless the propagation direction changes, i.e., it is in a single plane containing the line of propagation. *Note:* If the time-varying electric vector can be broken into two perpendicular components that have equal amplitudes and that differ in phase by $\frac{1}{4}$ wavelength, the light is said to be circularly polarized. Circular polarization is obtained whenever the phase differences between the two perpendicular components is any odd, integral number of quarter wavelengths. If the electric vector is resolvable into two perpendicular components of unlike amplitudes and differing in phase by values other than $\frac{1}{4}, \frac{1}{2}, \frac{3}{4}$, 1, etc., wavelengths, the light beam is said to be elliptically polarized.

polarizer An optical device capable of transforming unpolarized, i.e., diffused or scattered light, into polarized light, or altering the polarization of polarized light. *See also: FILTER.*

polychromatic radiation Electromagnetic radiation consisting of two or more frequencies or wavelengths. *See also: MONOCHROMATIC RADIATION.*

polychromatism *See: DICHROISM.*

population inversion 1. A redistribution of energy levels in a population of elements that is opposite to the equilibrium situation of having more atoms with lower energy-level electrons than atoms with higher energy-level electrons, i.e., an increase in the total number of electrons in the higher excited states occurs at the expense of the energy in the electrons in the ground or lower state and at the expense of the resonant energy source, i.e., the pump. This is not an equilibrium condition. The generation of population inversion is caused by pumping. 2. A condition in a stimulated material, such as a semiconductor, in which the upper energy level or band of two possible electronic energy levels or bands in a given atom, distribution of atoms, molecule, or distribution of molecules, has a higher probability, usually only slightly higher but nevertheless higher, of being occupied by an electron. *Note:* When population inversion occurs, the probability of downward energy transition giving rise to radiation, is greater than the probability of upward energy transitions, thus allowing light amplification by successive photon-triggered light emissions; i.e., stimulated emission.

position modulation *See: PULSE-POSITION MODULATION.*

positive feedback *Synonym: REGENERATION.*

positive-index guide A waveconfining dielectric structure for which the index of refraction of the central part is higher than the indices of refraction of the outer parts. *See also: WAVEGUIDE.*

positive-intrinsic-negative photodiode coupler A coupling device that enables the coupling of light energy from an optical fiber or cable onto the photo sensitive surface of a Positive-Intrinsic-Negative (PIN) diode of a photon detector (photodetector) at the receiving end of an optical fiber data link. *Note:* The coupler may only be a fiber pigtail epoxied to the photo diode.

positive justification *Synonym: BIT STUFFING.*

positive lens *See: CONVERGING LENS.*

post-detection combiner A circuit or device for combining two or more signals after demodulation. *See also: DIVERSITY COMBINER.*

power *See: IRRADIANCE; RADIANT INTENSITY; RADIANT POWER.*

power density *Colloquial synonym for IR-RADIANCE.*

power efficiency *See: OPTICAL POWER EFFICIENCY.*

power gain of an antenna The ratio of the power required at the input of a reference antenna to the power supplied to the input of the given antenna to produce, in a given direction, the same field at the same distance.

Note: When not specified otherwise, the figure expressing the gain of an antenna refers to the gain in the direction of the radiation main lobe. In services using scattering modes of propagation, the full gain of an antenna may not be realizable in practice and the apparent gain may vary with time. *Synonym: ANTENNA GAIN. See also: DIRECTIVE GAIN.*

power-law index profile A class of graded index profiles characterized by the following equations:

$$n(r) = n_1(1 - 2\Delta(r/a)^g)^{1/2} \qquad r \leq a$$

$$n(r) = n_2 = n_1(1 - 2\Delta)^{1/2} \qquad r \geq a$$

$$\text{where } \Delta = \frac{n_1^2 - n_2^2}{2n_1^2}$$

where $n(r)$ is the refractive index as a function of radius, n_1 is the refractive index on axis, n_2 is the refractive index of the homogeneous cladding, a is the core radius, and g is a parameter that defines the shape of the profile.

Note: 1. α is often used in place of g. Hence, this is sometimes called an alpha profile.

Note: 2. For this class of profiles, multimode distortion is smallest when g takes a particular value depending on the material used. For most materials, this optimum value is around 2. When g increases without limit, the profile tends to a step index profile. *See also: GRADED INDEX PROFILE; MODE VOLUME; PROFILE PARAMETER; STEP INDEX PROFILE.*

power level At any point in a transmission system, the ratio of the power at that point to some arbitrary amount of power chosen as a reference. This ratio is usually expressed either in decibels referred to one milliwatt, abbreviated dBm, or in decibels referred to one watt, abbreviated dBW. *See also: POWER.*

power points *See: EMISSION-BEAM-ANGLE BETWEEN HALF-POWER-POINTS.*

power ratio *See: RADIANT POWER RATIO.*

ppb *Acronym for PARTS PER BILLION.*

ppm *Acronym for PARTS PER MILLION.*

PPM *Acronym for PULSE-POSITION MODULATION.*

precipitation attenuation The loss in electromagnetic energy by scattering, refraction, and absorption during passage through a volume of the atmosphere containing precipitation such as rain, snow, hail, or sleet.

precipitation static A type of interference experienced in a receiver during snowstorms, rainstorms, and duststorms. It is caused by the impact of charged particles against an antenna. *See also: NOISE.*

precision-sleeve splicer A round tube, with a round hole that has a diameter equal to the outer diameter of two optical fibers to be spliced, containing a matched-index epoxy, into which the two fibers may be inserted from opposite ends.

Note: The ends of the sleeve may be crimped to hold the fibers tightly while the epoxy cures. *See also: LOOSE-TUBE SPLICER; TANGENTIAL COUPLING.*

predetection combiner A circuit or device for combining two or more signals prior to demodulation. *See also: DIVERSITY COMBINER.*

predetection combining A technique used to obtain an improved signal from multiple radio receivers involved in diversity reception.

Note: This process requires that all incoming diversity signals be brought into approximate phase coincidence before combining.

preemphasis A process in a system designed to increase the magnitude of some frequency components with respect to the magnitude of others in order to reduce adverse effects such as noise in subsequent parts of the system.

preemphasis improvement The improvement in the signal-to-noise ratio of the high-frequency end of the baseband resulting from passing the modulating signal (at the transmitter) through a preemphasis network which increases the magnitude of the higher signal frequencies, and then passing the output of the discriminator through a deemphasis network to restore the original signal power distribution.

preemphasis network A network inserted in a system in order to increase the magnitude of one range of frequencies with respect to another.

Note: Pre-emphasis is usually employed in FM or PM transmitters to equalize the modulating signal drive power in terms of deviation ratio. The receiver demodulation process includes a reciprocal network called de-emphasis to restore the original signal power distribution. *See also: DE-EMPHASIS.*

preform A glass structure from which an optical fiber waveguide may be drawn. *See also: CHEMICAL VAPOR DEPOSITION TECHNIQUE; ION EXCHANGE TECHNIQUE; OPTICAL BLANK.*

Prentice's rule A means of determining prism power at any point on a lens.
Note: Prism power equals dioptric power multiplied by the distance in centimeters from the optical center.

primary coating The material in intimate contact with the cladding surface, applied to preserve the integrity of that surface. *See also: CLADDING.*

primary spectrum The main, first, or the characteristic chromatic aberration of a simple nonachromatized lens or prism. *See also: SECONDARY SPECTRUM.*

principal focus *See: FOCAL POINT; PRINCIPAL FOCUS POINT.*

principal focus point The point to which incident parallel rays of light converge, or from which they diverge when they have been acted upon by a lens or mirror. A lens has a single point of principal focus on each side of the lens. A mirror has but one principal focus. A lens or mirror has an infinite number of image points, real or virtual, one for each position of the object. *Synonym: PRINCIPAL FOCUS.*

principal ray In the object space of an optical system, the ray directed at the first principal point, and hence in the image space, the ray, projected backward, intersecting the axis at the second principal point.

principle *See: FERMAT PRINCIPLE.*

prism A transparent body with at least two polished plane faces inclined with respect to each other, from which light is reflected or through which light is refracted.
Note: When light is refracted by a prism whose refractive index exceeds that of the surrounding medium, it is deviated or bent toward the thicker part of the prism. *See also: PENTA PRISM.*

prism chromatic resolving power When parallel rays of light are incident on a prism, the prism is oriented at the angle of minimum deviation at wavelength L, and the entire height of the prism is utilized, the resolving power, R, deduced on the basis of Rayleigh's criterion,

$$R = \frac{L}{\Delta L} = B\left(\frac{DN}{DL}\right)$$

where N is the index of refraction of the prism for the wavelength L and B is the maximum thickness of prism traversed by the light rays.
Note: The quantities DN/DL and B are often called the dispersion and baselength of the prism, respectively.

prismograph A graphic device for determining prism power.

probe *See: FIBER-OPTIC PROBE.*

process *See: AXIAL VAPOR-PHASE OXIDATION PROCESS; DOUBLE-CRUCIBLE PROCESS; MODIFIED INSIDE-VAPOR PHASE OXIDATION PROCESS; INSIDE VAPOR-PHASE OXIDATION PROCESS; CHEMICAL VAPOR-PHASE OXIDATION PROCESS; MODIFIED CHEMICAL-VAPOR DEPOSITION PROCESS; MOLECULAR STUFFING PROCESS; OUTSIDE VAPOR-PHASE OXIDATION PROCESS; PLASMA-ACTIVATED CHEMICAL VAPOR-DEPOSITION PROCESS.*

product *See: BIT-RATE-LENGTH PRODUCT; GAIN-BANDWIDTH PRODUCT.*

profile *See: GRADED INDEX PROFILE; INDEX PROFILE; PARABOLIC PROFILE; POWER-LAW INDEX PROFILE; STEP INDEX PROFILE.*

profile dispersion 1. In an optical waveguide, that dispersion attributable to the variation of refractive index contrast with wavelength, where contrast refers to the difference between the maximum refractive index in the core and the refractive index of the homogeneous cladding. Profile dispersion is usually characterized by the profile dispersion parameter, defined by the following entry. 2. In an optical waveguide, that dispersion attributable to the variation of refractive index profile with wavelength. The profile variation has two contributors: (a) variation in refractive index contrast, and (b) variation in profile parameter. *See also: DISPERSION; DISTORTION; REFRACTIVE INDEX PROFILE.*

profile dispersion parameter (P)

$$P(\lambda) = \frac{n_1}{N_1} \frac{\lambda}{\Delta} \frac{d\Delta}{d\lambda}$$

where n_1, N_1 are, respectively, the refractive and group indices of the core, and $n_1 \sqrt{1 - 2\Delta}$ is the refractive index of the homogeneous cladding, $N_1 = n_1 - \lambda(dn_1/d\lambda)$, and Δ is the refractive index constant. Sometimes it is defined with the factor (–2) in the numerator. *See also: DISPERSION.*

profile fiber *See: UNIFORM-INDEX-PROFILE FIBER.*

profile mismatch loss *See: REFRACTIVE-INDEX PROFILE MISMATCH LOSS.*

profile parameter The shape-defining parameter, g, for a power-law index profile. *See also: POWER-LAW INDEX PROFILE; REFRACTIVE INDEX PROFILE.*

programmer 1. That part of digital apparatus having the function of controlling the timing and sequencing of operations. 2. A person who prepares sequences of instructions for a computer.

program patch A temporary change in a computer program, routine, or subroutine. *Synonym: COMPUTER PATCH.*

propagation The motion of waves through or along a medium. *See also: ANOMALOUS PROPAGATION; DIFFRACTION; FORWARD SCATTER; IONOSPHERIC SCATTER; LINE-OF-SIGHT PROPAGATION; MULTIPATH; REFRACTION; SPORADIC E PROPAGATION; TROPOSPHERIC SCATTER.*

propagation constant For an electromagnetic field mode varying sinusoidally with time at a given frequency, the logarithmic rate of change, with respect to distance in a given direction, of the complex amplitude of any field component. *Note:* The propagation constant is a complex quantity.

propagation mode 1. The manner in which radio signals travel from a transmitting antenna to a receiving antenna, such as ground wave, sky wave, direct wave, ground reflection, or scatter. 2. One of the electric and magnetic field configurations in which energy propagates in a waveguide or along a transmission line. *See also: MODAL LOSS; MODE VOLUME.*

propagation path obstruction A man-made or natural physical feature that lies near enough to a radio path to cause a sensible effect on path loss, exclusive of reflection effects. *Note:* An obstruction may lie to the side, or even above the path, although usually it will lie below the path. Ridges, cliffs, buildings, and trees are examples of obstructions. If the clearance from the nearest anticipated path position, over the expected range of earth radius K-factor, exceeds 0.6 of the first Fresnel zone radius, the feature is not normally considered an obstruction. (NCS) *See also: EFFECTIVE EARTH RADIUS; FRESNEL ZONE; K-FACTOR; PATH CLEARANCE; PATH PROFILE.*

propagation time delay The time required for a signal to travel from one point to another.

proration 1. The distribution or allocation of parameters such as noise power proportionally among a number of tandem connected items, such as equipments, links, or trunks, in order to balance the performance of communications circuits. 2. In a telephone switching center, the distribution or allocation of equipments or components proportionally among a number of functions in order to provide a requisite grade of service.

protective coating *See: OPTICAL PROTECTIVE COATING.*

protective housing *See: LASER PROTECTIVE HOUSING.*

protocol The rules for communication system operation that must be followed if communication is to be effected. *Note:* Protocols may govern portions of a network, types of service, or administrative procedures. For example, a data link protocol is the specification of methods whereby data communication over a data link is performed in terms of the particular transmission mode, control procedures, and recovery procedures. *See also: LINK; NETWORK.*

pseudorandom noise Noise which satisfies one or more of the standard tests for statistical randomness. *Note:* Although it seems to lack any definite pattern, there is a sequence of pulses which repeats after a very long time interval.

PSK *Acronym for PHASE SHIFT KEYING.*

psophometer An instrument arranged to give visual indication corresponding to the aural effect of disturbing voltages of various frequencies.

Note: A psophometer usually incorporates a weighting network, the characteristics of which differ according to the type of circuit under consideration, e.g., high-quality music or commercial speech circuits.

psophometrically weighted dBm *See: dBm(psoph); dBmOp.*

psophometric voltage Circuit noise voltage measured in a line with a psophometer which includes a CCIF-1951 weighting network.

Note: 1. Do not confuse with psophometric emf, conceived as the emf in a generator (or line) with 600 ohms internal resistance, and hence, for practical purposes, numerically double the corresponding psophometric voltage.

Note: 2. Psophometric voltage readings, v (in millivolts), are commonly converted to dBm(psoph) by the relation:

$$dBm(psoph) = 20 \log_{10} v - 57.78 \cdot$$

See also: NOISE WEIGHTING.

psophometric weighting A noise weighting established by the International Consultative Committee for Telephony (CCIF, now CCITT), designated as CCIF-1951 weighting, for use in a noise measuring set or psophometer.

Note: The shape of this characteristic is virtually identical to that of F1A weighting. The psophometer is, however, calibrated with a tone of 800 Hz, 0 dBm, so that the corresponding voltage across 600 ohms produces a reading of 0.775 V. This introduces a 1 dB adjustment in the formulas for conversion with dBa. *See also: dBm(psoph).*

PTM *Acronym for PULSE TIME MODULATION.*

p-type semiconductor material Semiconductor material for which the dopant(s) create excess holes. The dopant atoms providing holes are called acceptors. In most stripe-geometry diode lasers the stripe contact is defined on the p-side. Most diode lasers are mounted with the p-side down on the heatsink (otherwise, the diodes are specified as "p-side up" devices).

pulse One of the elements of a repetitive signal characterized by the rise and decay in time of its magnitude, and usually short in relation to the time span of interest. *See also: ELECTROMAGNETIC PULSE; GAUSSIAN-SHAPED PULSE; IMPULSE; PULSE AMPLITUDE; PULSE DECAY TIME; PULSE DURATION; PULSE REPETITION FREQUENCY; PULSE RISE TIME.*

pulse-address multiple access The ability of a communication satellite to receive signals from several earth terminals simultaneously and to amplify, translate, and relay the signals back to earth, based on the addressing of each station by an assignment of a unique combination of time and frequency slots.

Note: This ability may be restricted by allowing only some of the terminals access to the satellite at any given time.

pulse amplitude The magnitude of a pulse.

Note: For a specific designation, adjectives such as average, instantaneous, peak, root-mean-square, etc., should be used to indicate the particular meaning intended. Pulse amplitude is measured with respect to a specified reference value. *See also: PULSE.*

pulse-amplitude modulation (PAM) That form of modulation in which the amplitude of the pulse carrier is varied in accordance with some characteristic of the modulating signal.

pulse broadening An increase in pulse duration.

Note: Pulse broadening may be specified by the impulse response, the root-mean-square pulse broadening, or the full-duration-half-maximum pulse broadening. *See also: IMPULSE RESPONSE; ROOT-MEAN-SQUARE PULSE BROADENING; FULL WIDTH (DURATION) HALF MAXIMUM.*

pulse-code modulation (PCM) That form of modulation in which the modulating signal is sampled, the sample quantized and coded, so that each element of information consists of different kinds of numbers of pulses and spaces. *See also: BALANCED CODE; CODE CONVERSION; DISPARITY; EQUIVALENT PCM NOISE; FRAME; LOAD CAPACITY; MULTIFRAME; PCM MULTIPLEX EQUIPMENT; PEAK LIMITING; SIGNALING.*

pulse decay time The time required for the pulse amplitude to go from 90 percent to 10 percent of the peak value. *Synonym: FALL TIME. See also: PULSE; PULSE RISE TIME.*

pulse dispersion A separation or spreading of input optical signals along the length of a transmission line, such as an optical fiber.

Note: This limits the useful transmission

bandwidth of the fiber. It's expressed in time and distance as nanoseconds per kilometer. Three basic mechanisms for dispersion are the material effect, the waveguide effect, and the multimode effect. Specific causes include surface roughness, presence of scattering centers, bends in the guiding structure, deformation of the guide and inhomogeneities of the guiding medium. *Synonym: PULSE SPREADING.*

pulse distortion *See: DISTORTION.*

pulsed laser *See: Q-SWITCHED REPETITIVELY PULSED LASER.*

pulse duration The time between a specified reference point on the first transition of a pulse waveform and a similarly specified point on the last transition. The time between the 10%, 50%, or 1/e points is commonly used, as is the rms pulse duration. *See also: ROOT-MEAN-SQUARE PULSE DURATION.*

pulse-duration modulation (PDM) That form of modulation in which the duration of a pulse is varied in accordance with some characteristic of the modulating signal. *Synonyms: PULSE-LENGTH MODULATION; PULSE WIDTH MODULATION.*

pulse duty factor The ratio of average pulse duration to average pulse spacing, therefore a dimensionless quantity.
Note: The spacing is time between pulses.

pulse-frequency modulation (PFM) That form of modulation in which the pulse repetition frequency of the carrier is varied in accordance with some characteristic of the modulating signal.

pulse length Often erroneously used as a synonym for Pulse duration.

pulse-length modulation *Synonym: PULSE-DURATION MODULATION.*

pulse-position modulation (PPM) 1. In an optical transmission system, modulation that causes the arrival time of pulses at a detector to vary according to a signal impressed on a pulsed source. The detector output being a function of

the arrival time with respect to a fixed reference. 2. That form of modulation in which the positions in time of the pulses are varied, in accordance with some characteristic of the modulating signals, without modifying the pulse width.

pulse repetition frequency In radar, the number of pulses that occur each second. Not to be confused with transmission frequency which is determined by the rate at which cycles are repeated within the transmitted pulse. *See also: PULSE.*

pulse rise time The time required for the pulse amplitude to go from 10 percent to 90 percent of the peak value. *See also: PULSE; PULSE DECAY TIME.*

pulse spreading *See: PULSE DISPERSION.*

pulse string *Synonym: PULSE TRAIN.*

pulse stuffing *Synonym: BIT STUFFING.*

pulse time modulation (PTM) Those forms of modulation in which the time of occurrence of some characteristics of the pulse carrier is varied with respect to some characteristic of the modulating signal.
Note: This includes pulse position modulation and pulse duration modulation.

pulse train A series of pulses having similar characteristics. *Synonym: PULSE STRING.*

pulse width Often erroneously used as a synonym for Pulse duration.

pulse-width modulation *Synonym: PULSE-DURATION MODULATION.*

pupil *See: ARTIFICIAL PUPIL; ENTRANCE PUPIL; EXIT PUPIL.*

pW *Acronym for PICOWATT.* A unit of power equal to 10^{-12}W (-90 dBm).
Note: It is commonly used for both weighted and unweighted noise measurements. Context must be observed. *See also: pWp.*

pWp0 Psophometrically weighted power in picowatts referred to as zero transmission level point (0TLP).

pWp, pW Psophometrically weighted. *See also: pW; NOISE WEIGHTING.*

QA *Acronym for* QUALITY ASSURANCE.
QC *Acronym for* QUALITY CONTROL.
QPSK *Acronym for* QUADRATURE PHASE SHIFT KEYING.

Q-switched repetitively-pulsed laser A solid-state laser whose continuous emission is converted into pulses by a Q-switch.

quadratic profile *Synonym:* PARABOLIC PROFILE.

quadrature phase shift keying (QPSK) PSK using four phase states. *Synonyms:* QUADRIPHASE; QUATERNARY PHASE SHIFT KEYING.

quadriphase *Synonym:* QUADRATURE PHASE SHIFT KEYING.

quadruple diversity The term applied to the simultaneous combining of, or selection from, four independently fading signals and their detection through the use of space, frequency, angle, time, or polarization characteristics or combinations thereof. *See also:* DIVERSITY RECEPTION; DUAL DIVERSITY; ORDER OF DIVERSITY.

quality *See:* IMAGE QUALITY.

quantity *See:* LIGHT QUANTITY.

quantization A process in which the continuous range of values of a signal is divided into nonoverlapping, but not necessarily equal subranges, and to each subrange a discrete value of the output is uniquely assigned. Whenever the signal value falls within a given subrange, the output has the corresponding discrete value. *See also:* QUANTIZATION LEVEL; SIGNAL; UNIFORM ENCODING.

quantization level The discrete value of the output designating a particular subrange of the input. *See also:* QUANTIZATION; UNIFORM ENCODING.

quantizing distortion The distortion resulting from the quantization process.

quantizing noise An undesirable random signal caused by the error of approximation in a quantizing process.
Note: It is solely dependent on the particular quantization process used and the statistical characteristics of the quantized signal.

quantum efficiency In an optical source or detector, the ratio of output quanta to input quanta. Input and output quanta need not both be photons.

quantum noise Noise atrributable to the discrete or particle nature of light. *Synonym:* PHOTON NOISE.

quantum-noise-limited operation Operation wherein the minimum detectable signal is limited by quantum noise. *See also:* QUANTUM NOISE.

quasi-analog signal A digital signal which has been converted into a form suitable for transmission over a specified analog channel.
Note: The specification of the analog channel would include frequency range, frequency bandwidth, S/N ratio, and envelope delay distortion. When this form of signaling is used to convey message traffic over dialed-up telephone systems, it is often referred to as voice-data. A modem may be used for the conversion process. *See also:* MODEM.

quaternary compound Semiconductor compound made of four elements. *See:* $In_xGa_{1-x}As_yP_{1-y}$.

quaternary devices Diode lasers, LEDs, or detectors for which the active medium is a quaternary compound. Generally used to describe long-wavelength InGaAsP lasers, although lasers of InGaAsP active layers on GaAs substrates are short-wavelength devices.

quaternary phase shift keying *Synonym:* QUADRATURE PHASE SHIFT KEYING.

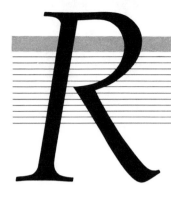

radial distortion An aberration of lens systems characterized by the imaging of an extra-axial straight line as a curved line without necessarily affecting the definition.

Note: Unsymmetrical, or otherwise irregular distortions of the image can also be caused by imperfect location of optical centers, or irregularity of optical surfaces.

radiance Radiant power, in a given direction, per unit solid angle per unit of projected area of the source, as viewed from that given direction. Radiance is expressed in watts per steradian per square meter. *See also: BRIGHTNESS; CONSERVATION OF RADIANCE; RADIOMETRY.*

radiance conservation The principle that states that passive optical paraphernalia cannot increase the radiance of a source, namely, the radiance of an image cannot exceed that of the object when energy is not added to the system. *Synonym: BRIGHTNESS CONSERVATION.*

radiance theorem *Synonym: CONSERVATION OF RADIANCE.*

radiant emittance Radiant power emitted into a full sphere (4π steradians) by a unit area of a source; expressed in watts per square meter. *Synonym: RADIANT EXITANCE. See also: RADIOMETRY.*

radiant energy Energy that is transferred via electromagnetic waves, i.e., the time integral of radiant power; expressed in joules. *See also: RADIOMETRY.*

radiant exitance *Synonym: RADIANT EMITTANCE.*

radiant flux *Synonym: RADIANT POWER (OBSOLETE).*

radiant incidence *See: IRRADIANCE.*

radiant intensity Radiant power per unit solid angle, expressed in watts per steradian. *See also: INTENSITY; RADIOMETRY.*

radiant power The time rate of flow of radiant energy, expressed in watts. The prefix is often dropped and the term "power" is used. *Colloquial synonyms: FLUX; OPTICAL POWER; POWER; RADIANT FLUX. See also: RADIOMETRY.*

radiant transmittance The ratio of the radiant flux transmitted by an object to the incident radiant flux.

radiation 1. The electromagnetic waves or photons emitted from a source. 2. In radio communication, the emission of energy in the form of electromagnetic waves. *See also: RADIATION PATTERN; MICROWAVE-AMPLIFICATION-BY-STIMULATED-EMISSION-OF-RADIATION; MONOCHROMATIC RADIATION; POLYCHROMATIC RADIATION; SPURIOUS RADIATION; THERMAL RADIATION.*

radiation angle Half the vertex angle of the cone of light emitted by a fiber.

Note: The cone is usually defined by the angle at which the far-field irradiance has decreased to a specified fraction of its maximum value or as the cone within which can be found a specified fraction of the total radiated power at any point in the far field. *Synonym: OUTPUT ANGLE. See also: ACCEPTANCE ANGLE; FAR-FIELD REGION; NUMERICAL APERTURE.*

radiation efficiency *See: LUMINOUS RADIATION EFFICIENCY.*

radiation field *Synonym: FAR FIELD REGION.*

radiation losses Optical losses due to transmission of light at the dielectric boundaries of an antiguide. *Synonym: LEAKAGE LOSSES.*

Note: Radiation losses in an antiguide are a strong function of the mode number, and thus filter out the high-order modes.

radiation mode In an optical waveguide, a mode whose fields are transversely oscillatory everywhere external to the waveguide, and which exists even in the limit of zero wavelength. Specifically, a mode for which

$$\beta \leq [n^2(a)k^2 - (l/a)^2]^{1/2}$$

where β is the imaginary part (phase term) of the axial propagation constant, l is the azimuthal index of the mode, $n(a)$ is the refractive index at $r = a$, the core radius, and k is the free-space wavenumber, $2\pi/\lambda$, where λ is the wavelength. Radiation modes correspond to refracted rays in the terminology of geometric optics. *Synonym: UNBOUND MODE. See also: BOUND MODE; LEAKY MODE; MODE; REFRACTED RAY.*

radiation pattern   Relative power distribution as a function of position or angle. *Note: 1.* Near-field radiation pattern describes the radiant emittance ($W \cdot m^{-2}$) as a function of position in the plane of the exit face of an optical fiber. *Note: 2.* Far-field radiation pattern describes the irradiance as a function of angle in the far field region of the exit face of an optical fiber. *Note: 3.* Radiation pattern may be a function of the length of the waveguide, the manner in which it is excited, and the wavelength. *See also: FAR-FIELD RE-GION; NEAR-FIELD REGION.*

radiation scattering   The diversion of radiation (thermal, electromagnetic, or nuclear) from its original path as a result of interactions or collisions with atoms, molecules, or large particles in the atmosphere or other media between the source of radiation (e.g., a nuclear explosion) and a point some distance away. As a result of scattering, radiation (especially gamma rays and neutrons) will be received at such a point from many directions instead of only from the direction of the source. *See also: ELECTRO-MAGNETIC RADIATION; RADIATION; SCATTER.*

radiation temperature *See: TOTAL RADIATION TEMPERATURE.*

radiative carrier lifetime   The average decay time for a given concentration of excess carriers that recombine radiatively. At low injection levels in edge-emitting LEDs, and at most injection levels in surface-emitting LEDs, the carrier lifetime is determined by the background carrier concentration. At high injection levels the radiative carrier lifetime is determined by the excess carrier concentration, up to lasing threshold. Spontaneous

carrier lifetimes have values in the 1–10 ns range. In stimulated emission above threshold the excess carrier lifetime is much shorter than in spontaneous emission. For devices of internal quantum efficiency close to 100% the radiative carrier lifetime determines the frequency response. *Note:* For examples see H. Kressel and J. K. Butler, *Semiconductor Lasers and Heterojunction LEDs,* Academic Press, 1977.

radiative recombination   In an electroluminescent diode in which electrons and holes are injected into the recombination region by application of a forward bias, the carrier recombination mechanism that produces photons. The emitted photons have an energy ($h\gamma$) approximately equal to the bandgap energy. *Note:* Radiative recombination produces the light in a LED or LD, which can be modulated for signal transmissions through fibers. *See also: NONRADIATIVE RECOMBINATION.*

radii loss *See: MISMATCH-OF-CORE-RADII LOSS.*

radio beam   A radiation pattern from a directional antenna such that the energy of the transmitted electromagnetic wave is confined to a small angle in at least one dimension. *See also: BEAMWIDTH.*

radio frequency interference (RFI) *Synonym: ELECTROMAGNETIC INTER-FERENCE.*

radio frequency (RF)   Those frequencies of the electromagnetic spectrum normally associated with radio wave propagation. *Note:* For designation of subdivisions, *see: SPECTRUM DESIGNATION OF FRE-QUENCY.*

radiometer   An instrument designed to measure radiant intensity.

radiometry   The science of radiation measurement. The basic quantities of radiometry are listed on page 130.

radius *See: CRITICAL RADIUS.*

Raman effect *See: RAMAN SCATTER-ING.*

Raman laser *See: STIMULATED RA-MAN SCATTERING.*

Raman scattering   The part of the scattered radiation in a medium (gas, liquid, or solid) whose frequency is shifted from the incident radiation by amounts corresponding to molecular frequencies of vibration or rotation. Raman spectra give "fingerprints" of the molecular struc-

RADIOMETRIC TERMS			
TERM NAME	SYMBOL	QUANTITY	UNIT
Radiant energy	Q	Energy	joule (J)
Radiant power Synonym: Optical power	φ	Power	watt (W)
Irradiance	E	Power incident per unit area (irrespective of angle)	W·m⁻²
Spectral irradiance	E_λ	Irradiance per unit wavelength interval at a given wavelength	W·m⁻² ·nm⁻¹
Radiant emittance Synomym: Radiant excitance	W	Power emitted (into a full sphere) per unit area	W·m⁻²
Radiant intensity	I	Power per unit solid angle	W·sr⁻¹
Radiance	L	Power per unit angle per unit projected area	W·sr⁻¹ ·m⁻²
Spectral radiance	L_λ	Radiance per unit wavelength interval at a given wavelength	W·sr⁻¹ ·m⁻²·nm⁻¹

ture. Of importance in optical fibers only at very high optical power densities, and only above a threshold (i.e., onset of stimulated Raman scattering). *Synonym: RAMAN EFFECT.*

randomizer A device used to invert the sense of pseudorandomly selected bits of a bit stream to avoid long sequences of bits of the same sense. The same selection pattern must be used on the receive terminal in order to restore the original bit stream.

random noise Noise consisting of a large number of transient disturbances with a statistically random distribution. *Note:* Thermal noise is a type of random noise.

RAPD *Acronym for REACH-THROUGH AVALANCHE PHOTODIODE.*

rate *See BIT ERROR RATE; BLOCK RATE EFFICIENCY; BLOCK TRANSFER RATE; DATA SIGNALING RATE; ERROR RATE; MAXIMUM USER SIGNALING RATE; MODULATION RATE; MULTI-*

PLEX AGGREGATE BIT RATE; SCANNING RATE.

rated output power That power available at a specified output of a device under specified conditions of operation. *Note:* This power may be further described, e.g., maximum rated output power, average rated output power.

rate of transmission *Synonym: EFFECTIVE SPEED OF TRANSMISSION.*

ratio *See: APERTURE RATIO; COMMON-MODE REJECTION RATIO; FRONT-TO-BACK RATIO; IN-BAND NOISE POWER RATIO; RADIANT POWER RATIO; SIGNAL-PLUS-NOISE TO NOISE RATIO; SIGNAL-TO-NOISE RATIO; SINGLE-SIDEBAND NOISE POWER RATIO; STANDING WAVE RATIO.*

ratio-squared combiner *Synonym: MAXIMAL-RATIO COMBINER.*

ray *See: LIGHT RAY.*

Rayleigh distribution A mathematical statement of the frequency distribution of random variables, for the case where the variables have the same variance and are not correlated.

Rayleigh fading Phase interference fading due to multipath which is approximated by the Rayleigh distribution. *See also: FADING; FADING DISTRIBUTION; MULTIPATH.*

Rayleigh scattering Light scattering by refractive index fluctuations (inhomogeneities in material density or composition) that are small with respect to wavelength. The scattering loss is inversely proportional to the fourth power of the wavelength. *See also: MATERIAL SCATTERING; SCATTERING; WAVEGUIDE SCATTERING.*

reach-through avalanche photodiode (RAPD) Avalance-photodiode structure for which virtually all photo-generated carriers are collected quickly. By contrast, in standard APDs part of the photocarriers are generated in undepleted material (e.g., substrate) from where collection can only be done by diffusion, and thus the device speed is limited. In RAPDs the speed of response is limited only by carrier transit times (100 ps to 1 ns). *Note:* For examples see H. W. Ruegg, *IEEE Trans. Electron. Devices* vol. ED-24, p. 239, 1967.

receiver *See: OPTICAL RECEIVER.*

receiving element The accepting terminus of a junction of optical elements.

recombination *See:* NONRADIATIVE RECOMBINATION; RADIATIVE RECOMBINATION.

recombination region In an opto-electronic device the region in which injected carriers recombine. In single-heterostructure devices the recombination region thickness is comparable to the carrier diffusion length. In double-heterostructure devices the recombination region is identical to the active region.

reduced carrier transmission *Synonym:* SUPPRESSED CARRIER TRANSMISSION.

REED *See:* RESTRICTED EDGE-EMITTING DIODE.

reference noise The magnitude of circuit noise that will produce a circuit noise-meter reading equal to that produced by 10^{-12} watt (-90 dBm) of electrical power at 1000 Hz for noise meters calibrated in dBrn(144-line) or dBrnC. For noise meters calibrated in dBa(F1A), the reference noise is adjusted to -85 dBm.

reference surface That surface of an optical fiber which is used to contact the transverse-alignment elements of a component such as a connector. For various fiber types, the reference surface might be the fiber core, cladding, or buffer layer surface.
Note: In certain cases the reference surface may not be an integral part of the fiber. *See also:* FERRULE; OPTICAL WAVEGUIDE CONNECTOR.

reflectance The ratio of reflected power to incident power.
Note: In optics, frequently expressed as optical density or as a percent; in communication applications, generally expressed in dB. Reflectance may be defined as specular or diffuse, depending on the nature of the reflecting surface. Formerly: "reflection." *See also:* REFLECTION.

reflectance loss *See:* FRESNEL REFLECTION LOSS.

reflected ray The ray of electromagnetic radiation, usually light, leaving a reflecting surface, representing its path after reflection.

reflection The abrupt change in direction of a light beam at an interface between two dissimilar media so that the light beam returns into the medium from which it originated. Reflection from a smooth surface is termed specular, whereas reflection from a rough surface is termed diffuse. *See also:* CRITICAL ANGLE; REFLECTANCE; REFLECTIVITY; TOTAL INTERNAL REFLECTION.

reflection angle When a ray of electromagnetic radiation strikes a surface, and is reflected in whole or in part by the surface, the angle between the normal total reflecting surface and the reflected ray. *See also:* CRITICAL ANGLE.

reflection coefficient 1. The ratio of the reflected field strength to the incident field strength when an electromagnetic wave is incident upon an interface surface between dielectric media of different indices of refraction. (If, at oblique incidence, the electric field component of the incident wave is parallel to the interface, the reflection coefficient is given by

$$R = \frac{N_2 \cos A - N_1 \cos B}{N_2 \cos A + N_1 \cos B}$$

where N_1 and N_2 are the indices of refraction of the incident and transmitted medium respectively, and A and B are the angles of incidence and refraction [with respect to normal] respectively. If, at oblique incidence, the magnetic field component of the incident wave is parallel to the interface, the reflection coefficient is given by

$$R = \frac{N_1 \cos A - N_2 \cos B}{N_1 \cos A + N_2 \cos B}$$

This is known as the Fresnel equation for these cases.) 2. In radio propagation, the ratio between the amplitude of the reflected wave and the amplitude of the incident wave. (For large smooth surfaces, the reflection coefficient may be near unity. At near grazing incidence, even rough surfaces may reflect relatively well.) 3. At any specified place in a transmission line between a source of power and an absorber of power, the vector ratio of the electric field associated with the reflected wave to that associated with the incident wave. The reflection coefficient, RC, is given by the formula:

$$RC = \frac{(Z_2 - Z_1)}{(Z_2 + Z_1)} = \frac{(SWR - 1)}{(SWR + 1)}$$

where Z_1 is the impedance toward the source, Z_2 is the impedance toward the load and SWR is the standing wave ratio. *See also:* TRANSMISSION COEFFICIENT; FRESNEL REFLECTION LOSS;

REFLECTION LOSS; RETURN LOSS; STANDING WAVE RATIO.

reflection image   An image formed by a reflecting surface.

Note: An unwanted reflection image is a ghost image.

reflection law   When a ray of electromagnetic radiation strikes a surface and is reflected in whole or in part by the surface, the angle of reflection is equal to the angle of incidence, the incident ray, reflected ray, and normal all being in the same plane.

reflective coating   *See: HIGHLY REFLECTIVE COATING.*

reflective star-coupler   An optical fiber coupling device that enables signals in one or more fibers to be transmitted to one or more other fibers by entering the signals into one side of an optical cylinder, fiber, or other piece of material, with a reflecting back surface so as to reflect the diffused signals back to the output ports on the same side of the material, for conduction away in one or more fibers. *See also: TEE COUPLER; NONREFLECTIVE STAR-COUPLER.*

reflectivity   The reflectance of the surface of a material so thick that the reflectance does not change with increasing thickness; the intrinsic reflectance of the surface, irrespective of other parameters such as the reflectance of the rear surface. No longer in common usage. *See also: REFLECTANCE.*

reflector   One or more conductors or conducting surfaces for reflecting radiant energy. *See also: TRIPLE MIRROR.*

refracted ray   In an optical waveguide, a ray that is refracted from the core into the cladding. Specifically a ray at radial position r having direction such that

$$\frac{n^2(r) - n^2(a)}{1 - (r/a)^2 \cos^2 \phi(r)} \leq \sin^2 \theta \ (r)$$

where $\phi(r)$ is the azimuthal angle of projection of the ray on the transverse plane, $\theta(r)$ is the angle the ray makes with the waveguide axis, $n(r)$ is the refractive index, $n(a)$ is the refractive index at the core radius, and a is the core radius. Refracted rays correspond to radiation modes in the terminology of mode descriptors. *See also: CLADDING RAY; GUIDED RAY; LEAKY RAY; RADIATION MODE.*

refracted ray method   The technique for measuring the index profile of an optical fiber by scanning the entrance face with the vertex of a high numerical aperture cone and measuring the change in power of refracted (unguided) rays. *Synonym: REFRACTED NEAR-FIELD SCANNING METHOD. See also: REFRACTION; REFRACTED RAY.*

refracting crystal   *See: DOUBLY REFRACTING CRYSTAL; MULTIREFRACTING CRYSTAL.*

refraction   The bending of a beam of light in transmission through an interface between two dissimilar media or in a medium whose refractive index is a continuous function of position (graded index medium). *See also: ANGLE OF DEVIATION; REFRACTIVE INDEX (OF A MEDIUM).*

refraction angle   When an electromagnetic wave strikes a surface and is wholly or partially transmitted into the new medium, of which the struck surface is the boundary, the acute angle between the normal to the refracting surface at the point of incidence, and the refracted ray.

refraction law   *See: SNELL'S LAW.*

refractive index   1. The ratio of the velocity of light in a vacuum to the velocity of light in the medium whose index of refraction is desired, for example, N = 2.6 for certain kinds of glass (Examples: vacuum, 1.000; air, 1.000292; water, 1.333; ordinary crown glass, 1.516; gallium arsenide, 3.6. Since the index of air is very close to that of vacuum, the two are often used interchangeably). 2. The ratio of the sines of the angle of incidence and the angle of refraction when light passes from one medium to another. The index between two media is the relative index, while the index when the first medium is a vacuum is the absolute index of the second medium. The index of refraction expressed in tables is the absolute index, that is, vacuum to substance at a certain temperature, with light of a certain wavelength. *Synonym: ABSOLUTE REFRACTIVE INDEX; INDEX-OF-REFRACTION. See also: CLADDING; CORE; CRITICAL ANGLE; DISPERSION; FRESNEL REFLECTION; FUSED SILICA; GRADED INDEX OPTICAL WAVEGUIDE; GROUP INDEX; INDEX MATCHING MATERIALS; INDEX PROFILE; LINEARLY POLARIZED MODE; MATERIAL DISPERSION; MODE; NORMALIZED FREQUENCY; NUMERICAL APERTURE; OPTICAL PATH LENGTH; POWER-LAW INDEX PROFILE; PRO-*

FILE DISPERSION; SCATTERING; STEP INDEX OPTICAL WAVEGUIDE; WEAKLY GUIDING OPTICAL WAVEGUIDE.

refractive index contrast Denoted by Δ, a measure of the relative difference in refractive index of the core and cladding of a fiber, given by

$$\Delta = (n_1^2 - n_2^2)/2n_1^2$$

where n_1 and n_2 are, respectively, the maximum refractive index in the core and the refractive index of the homogeneous cladding.

refractive index profile The description of the refractive index along a fiber diameter. *See also: GRADED INDEX PROFILE; PARABOLIC PROFILE; POWER-LAW INDEX PROFILE; PROFILE DISPERSION; PROFILE DISPERSION PARAMETER; PROFILE PARAMETER; STEP INDEX PROFILE.*

refractive-index-profile mismatch loss A loss of signal power introduced by an optical fiber splice of two optical fibers whose graded indices of refraction are not the same.

refracted near-field scanning method *See: REFRACTED RAY METHOD.*

regeneration 1. The gain that results from coupling the output of an amplifier to its input. *Synonym: POSITIVE FEEDBACK.* 2. The action of a regenerative repeater in which digital signals are amplified, reshaped, retimed and retransmitted. 3. In a storage device whose information storing state may deteriorate, the process of restoring the device to its latest undeteriorated state. *See also: SIGNAL REGENERATION.*

regenerative repeater A repeater that is designed for digital transmission. *Synonym: REGENERATOR. See also: OPTICAL REPEATER.*

regenerator *Synonym: REGENERATIVE REPEATER.*

relaxation oscillations Damped oscillations in the diode lasers' light output response to current pulses. Relaxation oscillations (RO) are a basic laser phenomenon in that they reflect instabilities in the degree of population inversion due to different photon and electron lifetimes. The oscillation amplitude varies with the structure and test conditions. At data rates above 100 Mbit/s RO can produce serious deterioration of the pulse shape. In mode-stabilized devices RO are appreciably damped and

thus allow operation up to and at GHz rates.

repeater A device which amplifies an input signal or, in the case of pulses, amplifies, reshapes, retimes, or performs a combination of any of these functions on an input signal for retransmission.
Note: It may be either a one-way or two-way type. *See also: BRANCHING REPEATER; BROADCAST REPEATER; CONFERENCE REPEATER; OPTICAL REPEATER; PASSIVE REPEATER; REGENERATIVE REPEATER.*

residual modulation *Synonym: CARRIER NOISE LEVEL.*

resolution angle *See: LIMITING RESOLUTION ANGLE.*

resolving power A measure of the ability of a lens or optical system to form separate and distinct images of two objects close together.
Note: Because of diffraction at the aperture, no optical system can form a perfect image of a point, but produces instead a small disk of light (airy disk) surrounded by alternately dark and bright concentric rings. When two object points are at that critical separation from which the first dark ring of one diffraction pattern falls upon the central disk of the other, the points are just resolved, i.e., distinguished as separated, and the points are said to be at the limit of resolution. *See also: CHROMATIC RESOLVING POWER; THEORETICAL RESOLVING POWER; BUNDLE RESOLVING POWER; GRATING CHROMATIC RESOLVING POWER; PRISM CHROMATIC RESOLVING POWER.*

resonant cavity *See: OPTICAL CAVITY.*

response quantum efficiency The ratio of the number of countable output events to the number of incident photons that occur when electromagnetic energy is incident upon a material, often measured as electrons emitted per incident photon.
Note: Response quantum efficiency is a measure of the efficiency of conversion or utilization of optical energy, being an indication of the number of events produced for each incident quantum for a photodetector, it is a measure of the probability that the photodetector triggers a measureable event when a photon is incident. Quantum efficiency is an intrinsic quality of materials, and a function of wavelength, angle of incidence and polarization of the incident electro-

magnetic field. Normally, it is the number of electrons released or emitted, on the average, for each incident photon, and can be determined experimentally. The creation of an electron-hole pair by an incident photon is a complex probabilistic phenomena that depends on the details of the energy band structure of the material.

responsivity The ratio of an optical detector's electrical output to its optical input, the precise definition depending on the detector type; generally expressed in amperes per watt or volts per watt of incident radiant power.
Note: "Sensitivity" is often incorrectly used as a synonym.

restricted edge-emitting diode (REED) An edge-emitting LED, i.e., a light-emitting diode in which light is emitted only over a small portion of an edge.
Note: The restricted light-emitting region improves coupling efficiency with optical fibers and integrated optical circuits.

reticle A scale, indicator, or pattern placed in one of the focal planes of an optical instrument that appears to the observer to be superimposed upon the field of view.
Note: Reticles, in various patterns, are used to determine the center of the field or to assist in the gaging of distance, determining leads, or measurement. A reticle may consist of fine wires, or fibers, mounted on a support at the ends, or may be etched on a clear, scrupulously polished and cleaned plane parallel plate of glass, in which case the entire piece of glass is the reticle.

retrodirective reflector *See: TRIPLE MIRROR.*

return loss The ratio, at the junction of a transmission line and a terminating impedance, of the amplitude of the reflected wave to the amplitude of the incident wave, expressed in dB.
Note: More broadly, the return loss is a measure of the dissimilarity between two impedances, being equal to the number of decibels which corresponds to the scalar value of the reciprocal of the reflection coefficient, RC, and hence being expressed by the formula:

$$\text{Return Loss} = 1/RC = (Z_2 + Z_1)/(Z_2 - Z_1)$$

where Z_1 is the impedance toward the source and Z_2 is the impedance toward

the load. *See also: LOSS; REFLECTION COEFFICIENT; REFLECTION LOSS.*

return-to-zero (RZ) code A code form having two information states called "zero" and "one", and having a third state or condition to which each signal returns during each period. *See also: NON-RETURN-TO-ZERO (NRZ) CODE.*

reverted In optical systems, turned the opposite way so that right becomes left, and vice versa, such as the effect produced by a mirror in reflecting an image.

reverted image In an optical system, an image, the right side of which appears to be the left side, and vice versa.

RF *Acronym for RADIO FREQUENCY.*

RF bandwidth The difference between the highest and the lowest emission frequencies, in the region of the carrier or principal carrier frequency.
Note: In practice, the region of the carrier or principal carrier frequency beyond which the amplitude of any frequency resulting from modulation by signal, subcarrier, or both frequencies and their distortion products is less than 5 percent (-26 dB) of the rated peak output amplitude of: (a) the carrier or a single-tone sideband, whichever is greater, for single-channel emission; or (b) any subcarrier or a single-tone sideband thereof, whichever is greater for multiplex emission.

RF power margin An extra amount of transmitter power that may be specified by a designer because of uncertainties in the empirical components of the prediction method, the terrain characteristics, atmospheric variability, and equipment performance parameters. *Synonym: DESIGN MARGIN. See also: FADE MARGIN.*

Richardson's law The basic law of thermionic emission, expressed by the Richardson Dushman Equation, i.e., the current density (amps/square meter) due to thermal excitation in the cathode material is,

$$J = AT^2 e^{(-BQ/kT)}$$

where T is the cathode temperature (absolute), k is Boltzmann's constant, and A is a constant. The theoretical value of A is 1.2×10 amps/square meter but departures from this value occur; Q is the electronic charge, B is the work function (joules/coulomb) for the cathode material and $k = 1.38 \times 10^{-23}$ joules/kelvin.

ridge waveguide In a diode laser a two-dimensional dielectric waveguide consisting of a local increase in active layer thickness in the lateral direction (in a cross-sectional plane). CDH and CNS lasers have ridge guides for lateral mode control.

rms pulse duration *See: ROOT-MEAN-SQUARE (rms) PULSE DURATION.*

RO *Acronym for RELAXATION OSCILLATIONS.*

rod coupler *See: FIBER-OPTIC ROD COUPLER.*

rod multiplexer-filter *See: FIBER-OPTIC ROD MULTIPLEXER-FILTER.*

roofing filter A low-pass filter used to reduce unwanted higher frequencies. *Synonym: ROOF FILTER.*

room noise level *Synonym: AMBIENT NOISE LEVEL.*

root-mean-square (rms) deviation A single quantity characterizing a function given, for f(x), by

$$\sigma_{rms} = [1/M_0 \int_{-\infty}^{\infty} (x - M_1)^2 \, f(x) \, dx]^{1/2}$$

where $M_0 = \int_{-\infty}^{\infty} f(x)dx$

$$M_1 = 1/M_0 \int_{-\infty}^{\infty} xf(x)dx$$

Note: The term rms deviation is also used in probability and statistics, where the normalization, M_0, is unity. Here, the term is used in a more general sense. *See also: IMPULSE RESPONSE; ROOT-MEAN-SQUARE (rms) PULSE BROADENING; ROOT-MEAN-SQUARE (rms) PULSE DURATION; SPECTRAL WIDTH.*

root-mean-square (rms) pulse broadening The temporal rms deviation of the impulse response of a system. *See: ROOT-MEAN-SQUARE (rms) DEVIATION; ROOT-MEAN-SQUARE (rms) PULSE DURATION.*

root-mean-square (rms) pulse duration A special case of root-mean-square deviation where the independent variable is time and f(t) is pulse waveform. *See also: ROOT-MEAN-SQUARE DEVIATION.*

rotation *See: OPTICAL ROTATION.*

rotator *See: IMAGE ROTATOR.*

rule *See: PRENTICE'S RULE.*

run *See: CABLE RUN.*

RZ *See: RETURN-TO-ZERO (RZ) CODE.*

SAM-APD *Acronym for SEPARATE-AB-SORPTION AND MULTIPLICATION-REGIONS APD.*

sampling frequency The rate at which signals in an individual channel are sampled for subsequent modulation, coding, quantization, or any combination of these functions. The sampling frequency is usually specified as the number of samples per unit time. *Synonym: SAMPLING RATE.*

sampling rate *Synonym: SAMPLING FREQUENCY.*

sampling time The reciprocal of the sampling frequency.

satellite An object or vehicle orbiting, or intended to orbit, the earth, moon, or other celestial body. *See also: ACTIVE SATELLITE; COMMUNICATIONS SATELLITE; DOWNLINK; EARTH COVERAGE; FOOTPRINT; SATELLITE CHANNEL MODES; SATELLITE EARTH TERMINAL; SATELLITE RELAY; SYNCHRONOUS ORBIT; SYNCHRONOUS SATELLITE; UPLINK; VIEW.*

satellite earth terminal That portion of a satellite link which receives, processes, and transmits communications between the earth and a satellite.

satellite relay An active or passive satellite repeater that relays signals between two earth terminals.

scatter The process whereby the direction, frequency or polarization of waves is changed when the waves encounter one or more discontinuities in the medium which have lengths on the order of a wavelength.
Note: The term is frequently used to imply a disordered change in the incident energy. *See also: BACK SCATTERING; FORWARD SCATTER; IONOSPHERIC SCATTER; RADIATION SCATTERING; TROPOSPHERIC SCATTER.*

scattering The change in direction of light rays or photons after striking a small particle or particles. It may also be regarded as the diffusion of a light beam caused by the inhomogeneity of the transmitting medium. *See also: LEAKY MODES; MATERIAL SCATTERING; MODE; NONLINEAR SCATTERING; RAYLEIGH SCATTERING; REFRACTIVE INDEX (OF A MEDIUM); UNBOUND MODE; WAVEGUIDE SCATTERING.*

scattering coefficient Density would produce noise which is the lower limit on detector noise; this leads to quantum-noise-limited sensitivity. The quantum limit to optical sensitivity is due to the granularity, or particle nature of light. Thus the minimum energy increment of an electromagnetic (optical) wave is hv, i.e., the energy of a photon, the noise of the photocurrent due to the optical signal. Thus, quantum noise becomes hvB in the limit, when photocurrent is due only to the optical signal, where h is Planck's constant, v is the frequency, and B is the bandwidth. *Synonym: QUANTUM NOISE.*

scattering loss Power loss by an electromagnetic wave due to random reflections and deflections of the waves caused by the material elements in the medium in which the waves are propagating as well as by impurities, imbedded particles, and inclusions.

scattered seed A few, occasional, easily visible coarse seeds.
Note: Several may be spaced 2 or 3 centimeters apart, but one here and there at much greater distance apart is more usual.

scintillation In radio propagation, a random fluctuation of the received field about its mean value, the deviations usually being relatively small.
Note: The effects of this phenomenon become more significant as frequency increases.

scope *See: FIBERSCOPE.*

scratch In optics, a marking or tearing of the surface appearing as though it had been done by either a sharp or rough instrument.

Note: Scratches occur on sheet glass in all degrees from various accidental causes. Block reek is a chain-like scratch produced in polishing. A runner-cut is a curved scratch caused by grinding. A sleek is a hairline scratch. A crush or rub is a surface scratch or series of small scratches generally caused by mishandling, scuffing, or scraping.

SDM *Acronym for SPACE DIVISION MULTIPLEX.*

secondary spectrum The residual chromatic aberration, particularly the longitudinal chromatic aberration of an achromatic lens.

Note: Unlike the primary spectrum, it causes the image formed in one particular color to lie nearest the lens, the images in all other colors being formed behind the first at distances that increase sharply towards both ends of the useful wavelength spectrum. *See also: PRIMARY SPECTRUM.*

secondary station In a data communication network, the station responsible for performing unbalanced link-level operations, as instructed by the primary station.

Note: A secondary station interprets received commands and generates responses. *See also: BACKWARD SUPERVISION; CONTROL STATION; DATA COMMUNICATION; MASTER STATION; PRIMARY STATION; SLAVE STATION; TRIBUTARY STATION.*

second-harmonic distortion Quantity characterizing small-signal sinusoidal modulation of diode lasers. The amount of the second harmonic with respect to the first harmonic is a measure of the linearity of the light output vs. drive current curve. For mode-stabilized devices second-harmonic distortion is between −40 dB and −60 dB below the fundamental level, depending on the device used and the peak amplitude of the signal.

security optical fiber *See: OPTICAL-FIBER TRAP.*

seed A gaseous inclusion having an extremely small diameter in glass or other transparent medium. *See also: SCATTERED SEED.*

seeding *See: HEAVY SEEDING.*

segmented encoding law An encoding law in which an approximation to a smooth law is obtained by a number of linear segments. *Synonym: PIECEWISE LINEAR ENCODING. See also: ENCODING LAW.*

selective absorption The act or process by which a substance absorbs, i.e., takes up, soaks up all the frequencies or colors contained in a beam of electromagnetic radiation, such as white light, except those that it reflects or transmits.

Note: Some substances are transparent to waves of certain frequencies, allowing them to be transmitted, while absorbing waves of other frequencies. Some reflecting surfaces will absorb light of certain frequencies and reflect others. The color of a transparent object is the color it transmits, and the color of an opaque object is the color it reflects. *See also: SELECTIVE TRANSMISSION.*

selective combiner A circuit or device for selecting one of two or more diversity signals in which only the signal having the most desirable characteristics is selected and used.

Note: The selection process may be designed to operate on signal amplitude, signal-to-noise ratio, transition characteristics, or other signal characteristics. *See also: DIVERSITY COMBINER; MAXIMAL-RATIO COMBINER.*

selective fading That type of fading in which the components of the received radio signal fluctuate independently. *See also: FADING.*

selective transmission The act or process by which a substance conducts or transmits all the colors or frequencies of a beam of white light, except those that it reflects or absorbs.

Note: Some substances transmit only certain colors and absorb or reflect all others. The color of a transparent object is the color it transmits. The color of an opaque object is the color it reflects. Absorbed colors are not seen. *See also: SELECTIVE ABSORPTION.*

selective transmittance The property of variation of transmittance with the wavelength of light transmitted through a substance.

SELFOC fiber Trade name for graded-index fiber of parabolic index profile. *See: PARABOLIC INDEX PROFILE.*

self-sustained oscillations Oscillatory phenomenon in the light output response of some diode lasers to current pulses. SSO are generally associated with aged and/or poor-quality devices, unnaturally high slope efficiencies in the light-current curves (i.e., "snap-on" characteristics) and excess noise. *Synonym: SELF-PULSATIONS.* Self-pulsing lasers can be used for generating short pulses.

semiconductor laser Laser for which the active medium is semiconducting material and the population inversion is obtained by carrier injection (i.e., injection laser diode), electron-beam, or optical pumping. *See: INJECTION LASER DIODE (ILD).*

semi-duplex operation A method of operation of a communications circuit wherein one end is duplex and the other end is simplex.

Note: Sometimes used in mobile systems with the base station duplex and the mobile station simplex. *See also: DUPLEX OPERATION; HALF-DUPLEX OPERATION; SIMPLEX OPERATION.*

semi-leaky waveguide Two-dimensional dielectric wave confining structure which has total internal reflection to one side (i.e., positive-index guiding) and partial reflection to the other side (i.e., antiguiding). Semi-leaky guides are obtained in CDH lasers while growing over misoriented substrates.

Note: For examples see D. Botez, *IEEE J. Quantum Electron.*, vol. 17, pp. 2290–2309, Dec. 1981.

sensitivity Imprecise synonym for Responsivity. In optical system receivers, the minimum power required to achieve a specified quality of performance in terms of output signal-to-noise ratio or other measure.

separate-absorption and multiplication-region APD III-V material APD for which the p-n junction is placed in high-bandgap material, in close proximity to low-bandgap light-absorbing material. The effect is that excess dark current due to tunneling is suppressed, and thus gain values in the thousands are possible. A typical example is an InGaAs/InP APD with the p-n junction in InP and $In_{0.53}Ga_{0.47}As$ material as light absorber. Symbol: SAM-APD.

Note: For examples see K. Nishida *et al.*, *Appl. Phys. Lett.*, vol. 35, pp. 251–253, 1979.

separated multiclad-layer (SML) diode laser Mode-stabilized diode laser fabricated by two-step liquid-phase epitaxy and stripe etching. Lateral mode control is provided by layers on both (lateral) sides of the lasing region, that are absorbing and/or antiguiding for the emitted light. These losses are larger for high-order modes than for the fundamental mode, and thus single-mode operation is achieved. The lossy layers are part of current-blocking back-biased regions that laterally confine the current. The SML structure has been used for 0.85 μm, 1.3 μm and 1.5 μm mode-stabilized lasers.

Note: See for example Ishikawa *et al.*, *IEEE J. Quantum Electron.*, vol. QE-17, 1226–1234, July 1981.

service connection *See: LASER SERVICE CONNECTION.*

SF *Acronym for SINGLE FREQUENCY.*

S-F *Acronym for STORE AND FORWARD; STORE AND FORWARD SWITCHING CENTER.*

shaped pulse *See: GAUSSIAN-SHAPED PULSE.*

short-wavelength diode lasers In optical communications semiconductor diode lasers emitting in the wavelength range below 1 μm. Typical short-wavelength devices are AlGaAs lasers which cover the 0.72–0.87 μm wavelength range.

shot noise Noise caused by current fluctuations due to the discrete nature of charge carriers and random and/or unpredictable emission of charged particles from an emitter.

Note: There is often a (minor) inconsistency in referring to shot noise in an optical system: many authors refer to shot noise loosely when speaking of the mean square shot noise current (amp²) rather than noise power (watts). *See also: QUANTUM NOISE.*

SID *Acronym for SUDDEN IONOSPHERIC DISTURBANCE.*

sidebands The spectral energy, distributed above and below a carrier, resulting from a modulation process. *See also: SINGLE-SIDEBAND SUPPRESSED CARRIER.*

sideband transmission That method of transmission in which frequencies produced by amplitude modulation occur above and below the carrier frequency. The frequencies above the carrier are called "upper sideband" and those below the carrier are called "lower sideband."

Note: 1. The two sidebands may carry the same or different information.

Note: 2. The carrier and either sideband may be suppressed independently.

Note: 3. In conventional AM, both sidebands carry the same information and the carrier is present. *See also: COMPATIBLE SIDEBAND TRANSMISSION; DOUBLE-SIDEBAND SUPPRESSED CARRIER TRANSMISSION; INDEPENDENT-SIDEBAND TRANSMISSION; SINGLE-SIDEBAND SUPPRESSED CARRIER; SINGLE SIDEBAND TRANSMISSION; SUPPRESSED CARRIER TRANSMISSION.*

signal 1. The intelligence, message, or control function to be conveyed over a communication system. 2. As applied to electronics, any transmitted electrical impulse. 3. Operationally, a type of message, the text of which consists of one or more letters, words, characters, signal flags, visual displays or special sounds, with prearranged meanings and which is conveyed or transmitted by visual, acoustical, or electrical means. *See also: ALTERNATE MARK INVERSION SIGNAL; ANALOG SIGNAL; BACKWARD SIGNAL; BLACK SIGNAL; BUNCHED FRAME-ALIGNMENT SIGNAL; CALL ACCEPTED SIGNAL; CALL CONTROL SIGNALS; CALL-NOT-ACCEPTED SIGNAL; CALL PROGRESS SIGNAL; CALLED-LINE IDENTIFICATION SIGNAL; CALLING-LINE IDENTIFICATION SIGNAL; CALLING SIGNAL; CHARACTER SIGNAL; CLEAR CONFIRMATION SIGNAL; CODE CONVERSION; CONNECTION-IN-PROGRESS SIGNAL; DIGITAL SIGNAL; DISTRIBUTED MODIFIED AMI; OCTET TIMING SIGNAL; PEAK SIGNAL LEVEL; PILOT; QUANTIZATION; QUASI-ANALOG SIGNAL; UNIFORM ENCODING.*

signal distance 1. The number of digit positions in which the corresponding digits of two binary words of the same length are different. 2. The number of digit positions in which the corresponding digits of two words of the same length in any radix are different. For example, the signal distance between 21415926 and 11475916 is 3. *Synonym: HAMMING DISTANCE.*

signaling data link *See: DATA LINK.*

signaling time slot A time slot starting at a particular phase or instant in each frame and allocated to the transmission of supervisory and control data. *See also: DATA.*

signal - plus - noise - to - noise ratio ((S+N)/N). The ratio of the amplitude of the desired signal plus the noise to the amplitude of the noise at a given point. This usually is expressed in dB. *See also: NOISE.*

signal regeneration The restoration, to the extent practical, of a signal to an original predetermined configuration, shape, or position in time or space.

Note: Signal regeneration may involve amplifying, clipping, clamping, differentiating, integrating, clocking, or other operations. *See also: REGENERATION.*

signal-to-noise ratio (SNR) The ratio of the amplitude of the desired signal to the amplitude of noise signals at a given point in time.

Note: Usually expressed in dB, and in terms of peak values for impulse noise and rms values for random noise. Both the signal and noise should be defined to avoid ambiguity, e.g., peak-signal to peak-noise ratio. *See also: CHANNEL NOISE LEVEL; NOISE.*

signal transition The change from one signaling condition to another, for example, the change from "mark" to "space" or from "space" to "mark". *See also: MARK; SPACE.*

significant condition of modulation A condition assumed by the appropriate device corresponding to the quantized vlaue(s) of the characteristic(s) chosen to form the modulation. *See also: MARK; RESTITUTION.*

silica cladded fiber *See: DOPED-SILICA CLADDED FIBER.*

silica fiber *See: LOW-LOSS FEP-CLAD SILICA FIBER; PLASTIC-CLAD SILICA FIBER.*

silica graded fiber *See: DOPED-SILICA GRADED FIBER.*

simple microscope *See: MAGNIFIER.*

simplexed circuit A two-wire circuit from which a simplex circuit is derived. The two-wire circuit and the simplex circuit may be used simultaneously. *See also: SIMPLEX CIRCUIT.*

simplex operation That type of operation which permits the transmission of signals in either direction alternately.

Note: In radio telegraph or data transmission systems, it may be either (a) the use of a single frequency, time slot, or code address for transmission, and another frequency, time slot, or code address for reception, or (b) the use of the same frequency, time slot, or code address for both transmission and reception. In wire telegraph systems, simplex operation may be employed over either a half-duplex circuit, or over a neutral direct current circuit. *See also: HALF-DUPLEX OPERATION; SEMI-DUPLEX OPERATION.*

simplex signaling (SX) Signaling using two conductors for a single channel; a center tapped coil or its equivalent is used at both ends for this purpose.
Note: The arrangement may be a one-way signaling scheme suitable for intra-office use, or the simplex legs may be connected to (full) duplex signaling circuits which then function like CX signaling with E&M lead control.

sine wave object An object having a sinusoidal variation of luminance, having the advantage that the image will have a sinusoidal variation of illuminance and the only effect of degeneration by a lens system will be to decrease the modulation in the image relative to that in the object.

sine wave response *See: MODULATION TRANSFER FUNCTION.*

singing point The threshold point at which additional gain in the system will cause self-oscillations.

single-bundle cable *See: SINGLE-CHANNEL SINGLE-BUNDLE CABLE.*

single-channel single-bundle cable A bundle of optical fibers with a protective covering.

single-channel single-fiber cable A single optical conductor usually with a protective covering.

single-ended synchronization A synchronization control method used between two locations in which phase error signals used to control the clock at one location are derived from a comparison of the phase of the incoming signals and the phase of the internal clock of the same location. *Synonym: SINGLE-ENDED CONTROL. See also: CLOCK; SYNCHRONIZATION.*

single fiber A discrete conductor of light waves.
Note: The discrete filament of optical material, glass or plastic is usually made with a lower-index cladding.

single-fiber cable *See: MULTICHANNEL SINGLE-FIBER CABLE; SINGLE-CHANNEL SINGLE-FIBER CABLE.*

single-fiber light guide *See: OPTICAL FIBER.*

single-frequency interference That interference caused by a single-frequency source; for example, interference in a data transmission line induced by a 60

Hz source. *See also: FREQUENCY; INTERFERENCE.*

single-frequency (SF) signaling In telephone communications, a method of conveying dialing or supervisory signals, or both, with one or more specified single frequencies.
Note: The signals are normally inband in long-haul communications.

single-frequency (SF) signaling system In telephone communications, a system that uses single-frequency signaling.
Note: The DCS transmits dc signaling pulses or supervisory signals, or both, over carrier channels or cable pairs on a 4-wire basis using a 2600 Hz signal tone. The conversion into tones, or vice versa, is done by SF signal units.

single-harmonic distortion The ratio of the power of any single harmonic frequency signal to the power of the fundamental frequency signal. This ratio is measured at the output of a device under specified conditions and is expressed in dB. *See also: HARMONIC DISTORTION; SECOND-HARMONIC DISTORTION; TOTAL HARMONIC DISTORTION.*

single heterojunction In a laser diode, a single junction involving two energy level shifts and two refractive index shifts, used to provide increased confinement of radiation direction, improved control of radiative recombination, and reduced nonradiative (thermal) recombination. *Synonym: CLOSE-CONFINEMENT JUNCTION.*

single-longitudinal-mode diode operation In diode lasers operation over wide ranges in drive current in a single-mode in frequency. The diode is then oscillating in a single spatial mode but not necessarily the fundamental one.

single-mode diode laser Diode laser that oscillates over wide ranges in drive current both in a single spatial mode as well as in a single mode in frequency. Mode-stabilized devices with built-in real-index lateral wave confinement usually oscillate in a single mode in cw operation at drive current levels above approximately 1.1 times the threshold current. However, for stable single-mode operation under modulation conditions external means such as injection locking, controlled reflection from an external mirror and/or grating structures have to be used.

single-mode fiber A fiber waveguide

that supports the propagation of only one mode.

Note: The single-mode fiber is usually a low-loss optical waveguide with a very small core (2-8 micrometers). It requires a laser source for the input signals because of the very small entrance aperture (acceptance cone). The small core radius approaches the wavelength of the source; consequently, only a single mode is propagated.

single-mode optical waveguide An optical waveguide in which only the lowest order bound mode (which may consist of a pair of orthogonally polarized fields) can propagate at the wavelength of interest. In step index guides, this occurs when the normalized frequency, V, is less than 2.405. For power-law profiles, single-mode operation occurs for normalized frequency, V, less than approximately 2.405 $\sqrt{(g+2)/g}$, where g is the profile parameter.

Note: In practice, the orthogonal polarizations may not be associated with degenerate modes. Synonym: *MONO-MODE OPTICAL WAVEGUIDE. See also: BOUND MODE; MODE; MULTIMODE OPTICAL WAVEGUIDE; NORMALIZED FREQUENCY; POWER-LAW INDEX PROFILE; PROFILE PARAMETER; STEP INDEX OPTICAL WAVEGUIDE.*

single-sideband equipment reference level The power of one of two equal tones which, when used together to modulate a transmitter, cause it to develop its full rated peak power output.

single-sideband noise power ratio The ratio of the power measured at the output, in the notch bandwidth, with the notch in, to the power in the notch bandwidth, with the notch out, again measured at the output, with the notch applied to an input sufficient to maintain the total system mean noise power output constant. *See also: NOISE.*

single-sideband suppressed carrier (SSB-SC) That method of single-sideband transmission wherein the carrier is suppressed. *See also: SIDEBAND TRANSMISSION; SIDEBANDS.*

single-sideband transmission That method of sideband transmission in which only one sideband is transmitted.

Note: The carrier may be suppressed.

single-spatial-mode diode operation In diode lasers of the narrow-stripe type or real-index laterally guided, operation over wide ranges in drive cur-rent in a single spatial mode. Usually the single mode is the fundamental mode (i.e., zeroth order mode) supported by the structure. Single-spatial-mode diode lasers need not necessarily oscillate in a single mode in frequency.

single-tone interference An undesired discrete frequency appearing in a signal channel.

skew ray A ray that does not intersect the optical axis of a system (in contrast with a meridional ray). *See also: AXIAL RAY; GEOMETRIC OPTICS; HYBRID MODE; MERIDIONAL RAY; OPTICAL AXIS; PARAXIAL RAY.*

skim In optical elements, streaks of dense seeds with accompanying small bubbles.

slab-dieletric optical waveguide An optical waveguide consisting of rectangular layers of ribbons of materials of differing refractive indices that support one or more lightwave transmission modes, with the energy of the transmitted waves confined primarily to the layer of highest refractive index, the lower indexed media serving as cladding, jacketing, or surrounding medium.

Note: Slab-dielectric optical waveguides are used in integrated optical circuits for geometrical convenience, in contrast to optical fibers in cables used for long-distance transmission.

slab interferometry The method for measuring the index profile of an optical fiber by preparing a thin sample that has its faces perpendicular to the axis of the fiber, and measuring its index profile by interferometry. *Synonym: AXIAL SLAB INTERFEROMETRY. See also: INTERFEROMETER.*

SLD *Acronym for SUPERLUMINESCENT DIODE.*

sleek In optical elements, a polishing scratch without visible conchoidal fracturing of the edges.

sleeve splicer *See: PRECISION-SLEEVE SPLICER.*

SML laser *See: SEPARATED-MULTI-CLAD LAYER DIODE LASER.*

smooth earth Idealized surfaces, such as water surfaces or very level terrain, having radio horizons that are not formed by prominent ridges or mountains but are determined solely as a function of antenna height above ground and the effective earth radius. *See also: PATH PROFILE.*

Snell's law When electromagnetic waves, such as light, pass from a given medium

141

to a denser medium, its path is deviated toward the normal: when passing into a less dense medium, its path is deviated away from normal.

Note: Snell's law, often called the law of refraction, defines this phenomenon by describing the relation between the angle of incidence and the angle of refraction as follows, namely,

$$\frac{\text{Sin I}}{\text{Sin R}} = \frac{N_R}{N_I}$$

where I is the angle of incidence, R is the angle of refraction, N_R is the refractive index of the medium containing the refracted ray, and N_I is the refractive index of the medium containing the incident ray. Stated in another way, both laws, that of reflection and of refraction, are attributed to Snell, namely, when the incident ray, the normal to the surface at the point of incidence of the ray on the surface, the reflected ray, and the refracted ray, all lie in a single plane. The angle between the incident ray and the normal is equal in magnitude to the angle between the reflected ray and the normal. The ratio of the sine of the angle between the normal and the incident ray is the sine of the angle between the normal and the refracted ray is a constant. *See also: REFRACTION.*

SNR *Acronym for SIGNAL-TO-NOISE RATIO.*

solid-state laser A laser whose active medium is a solid material such as glass, crystal, or semiconductor material, rather than gas or liquid. The term is generally used for glass lasers (e.g., Nd lasers) and ionic crystal lasers (e.g., ruby lasers).

source-coupler loss In an optical data link, optical communication system, or optical fiber system, the loss, usually expressed in dB, between the light source and the device or material that couples the light source energy from the source to the fiber cable.

source efficiency The ratio of emitted optical power of a source to the input electrical power.

source-fiber coupling In fiber optic transmission systems, the transfer of optical signal power emitted by a light source into an optical fiber, such coupling being dependent upon many factors, including geometry and fiber characteristics.

Note: Many optical fiber sources have an

optical fiber pigtail for connection by means of a splice or a connector to a transmission fiber.

source-to-fiber loss In an optical fiber, signal power loss caused by the distance of separation between a signal source and the conducting fiber.

space In binary modulation, the significant condition of modulation that is not specified as the "mark". *Synonyms: SPACING PULSE; SPACING SIGNAL. See also: MARK; MODULATION; SIGNAL TRANSITION.*

space-coherent light Light that has the property that over a given area, usually an area in a plane perpendicular to the direction of propagation, the amplitude, phase, and time variation are predictable and correlated.

Note: Spatial noncoherence refers to a random and unpredictable state of the phase over an area normal to the direction of propagation. *See also: COHERENT LIGHT; TIME-COHERENT LIGHT.*

space diversity A method of transmission or reception, or both, employed to minimize the effects of fading by the simultaneous use of two or more antennas spaced a number of wavelengths apart. *See also: DIVERSITY RECEPTION.*

space-division multiplexing *See: OPTICAL SPACE-DIVISION MULTIPLEXING.*

space-division switching A method whereby single transmission path routing determination is accomplished in a switch utilizing a physically separated set of matrix contacts or crosspoints. *See also: SWITCHING SYSTEM.*

spacing bias The uniform lengthening of all spacing signal pulses at the expense of all marking signal pulses. *See also: BIAS; MARKING BIAS.*

spacing pulse *Synonym: SPACE.*
spacing signal *Synonym: SPACE.*
spar *See: ICELAND SPAR.*
spatial coherence *See: COHERENT.*
spatially aligned bundle *See: ALIGNED BUNDLE.*
spatially coherent radiation *See: COHERENT.*

spatter Small chunks of material that fly from the hot crucible onto the glass surface, and adhere there, in evaporative coatings of optical elements such as lenses, prisms, and mirrors.

specification A document intended primarily for use in procurement, which clearly and accurately describes the essential technical requirements for items, materials, or services, including the procedures by which it will be determined that the requirements have been met. Specifications for items and materials may also contain preservation, packaging, packing, and marking requirements. *See also: DESIGN OBJECTIVE.*

specific detectivity   *Synonym: D*.*

speckle noise   *Synonym: MODAL NOISE.*

speckle pattern   A power intensity pattern produced by the mutual interference of partially coherent beams that are subject to minute temporal and spatial fluctuations.

Note: In a multimode fiber, a speckle pattern results from a superposition of mode field patterns. If the relative modal group velocities change with time, the speckle pattern will also change with time. If, in addition, differential mode attenuation is experienced, modal noise results. *See also: MODAL NOISE.*

spectral absorptance   The absorptance of electromagnetic radiation by a material evaluated at one or more wavelengths.

Note: Spectral absorptance is numerically the same for radiant and luminous flux.

spectral bandwidth   The wavelength interval in which a radiated spectral quantity is a specified fraction of its maximum value.

Note: The fraction is usually taken as 0.50 of the maximum power level, or 0.707 of the maximum (3 dB) current to voltage level. If the electromagnetic radiation is light, it is the radiant intensity half-power points that are used.

spectral density   The power density of electromagnetic radiation consisting of a continuous spectrum of frequencies, expressed in watts per hertz, taken over a finite bandwidth.

spectral emittance   The radiant emittance plotted as a function of wavelength.

spectral irradiance   Irradiance per unit wavelength interval at a given wavelength, expressed in watts per unit area per unit wavelength interval. *See also: IRRADIANCE; RADIOMETRY.*

spectral line   A narrow range of emitted or absorbed wavelengths. *See also: LINE SOURCE; LINE SPECTRUM; MONOCHROMATIC; SPECTRAL WIDTH.*

spectral order   *See: DIFFRACTION GRATING SPECTRAL ORDER.*

spectral radiance   Radiance per unit wavelength interval at a given wavelength, expressed in watts per steradian per unit area per wavelength interval. *See also: RADIANCE; RADIOMETRY.*

spectral reflectivity   The reflectivity of a surface evaluated as a function of wavelength.

spectral responsivity   Responsivity per unit wavelength interval at a given wavelength. *See also: RESPONSIVITY.*

spectral transmittance   Transmittance evaluated at one or more wavelengths, being numerically the same for radiant and luminous flux.

spectral width   A measure of the wavelength extent of a spectrum.

Note: 1. One method of specifying the spectral linewidth is the full width at half maximum (FWHM), specifically the difference between the wavelengths at which the magnitude drops to one-half of its maximum value. This method may be difficult to apply when the line has a complex shape.

Note: 2. Another method of specifying spectral width is a special case of root-mean-square deviation where the independent variable is wavelength (λ), and $f(\lambda)$ is a suitable radiometric quantity. *See also: ROOT-MEAN-SQUARE (RMS) DEVIATION.*

Note: 3. The relative spectral width $(\Delta\lambda)/\lambda$ is frequently used, where $\Delta\lambda$ is obtained according to Note 1 or Note 2. *See also: COHERENCE LENGTH; LINE SPECTRUM; MATERIAL DISPERSION.*

spectral window   A wavelength region of relatively high transmittance, surrounded by regions of low transmittance. *Synonym: TRANSMISSION WINDOW.*

spectrometer   A spectroscope provided with an angle scale capable of measuring the angular deviation of radiation of different wavelengths.

Note: In common usage, the dispersing means may be dispensed with, and the instrument used for measuring angles as on or through a prism.

spectroscope   An instrument capable of dispersing radiation into its component wavelengths and observing, or measuring, the resultant spectrum.

spectrum   *See: OPTICAL SPECTRUM.*

spectrum designation of frequency   A method of referring to a range or band of communication frequencies. In American practice the designator is a two or three letter abbreviation for the name. In ITU practice the designator is numeric. *See table below.*

spectrum signature The pattern of radio signal frequencies, amplitudes, and phases, which characterizes the output of a particular device, and tends to distinguish it from other devices.

specular reflection *See: REFLECTION.*

speech-plus-duplex operation That method of operation in which speech and telegraphy (duplex or simplex) are transmitted simultaneously over the same circuit, being kept from mutual interference by use of filters. *See also: COMPOSITED CIRCUIT.*

speech-plus signaling or telegraph An arrangement of equipment that permits the use of part of a speech band for transmission of signaling or telegraph signals. *See also: COMPOSITED CIRCUIT; SIGNALING.*

speed *See: LENS SPEED.*

spherical intensity *See: MEAN SPHERICAL INTENSITY.*

spherometer An instrument for the precise measurement of the radius of curvature of surfaces.

splice *See: OPTICAL WAVEGUIDE SPLICE.*

splice loss *See: INSERTION LOSS.*

splicer *See: LOOSE-TUBE SPLICER; PRECISION-SLEEVE SPLICER.*

splicing *See: FUSION SPLICING.*

spontaneous emission Light emitted randomly from a photoluminescent material when electrons excited to or injected at higher energy levels recombine radiatively.

Note: Spontaneous emission dominates the output from LEDs. In lasers sponta-

neous emission is significant up to and close to threshold. *See also: INJECTION LASER DIODE; LIGHT-EMITTING DIODE; RADIATIVE CARRIER LIFETIME; STIMULATED EMISSION; SUPER-RADIANCE.*

spontaneous emission efficiency *See: INTERNAL QUANTUM EFFICIENCY.*

spot size *See: LASING SPOT SIZE.*

spread *See: MULTIMODE GROUP-DELAY SPREAD.*

spreading *See: PULSE DISPERSION.*

spread spectrum 1. A communications technique in which the modulated information is transmitted in a bandwidth considerably greater than the frequency content of the original information.

Note: This technique affords advantages in interference avoidance and multiple access.

2. In general, pertaining to any signal with a large time-bandwidth product.

Note: In communications, spread spectrum techniques may be employed as an antinoise signal-gain processing tool. Spread spectrum pulses can be used for pulse addressing systems; other types of spread spectrum signals are wide-deviation FM with phase lock loop or frequency compressive feedback demodulators, frequency hopping, etc. Pseudorandom coding (numbers) are frequently employed as part of the "spread" process. *See also: FREQUENCY HOPPING; PSEUDORANDOM NUMBER SEQUENCE.*

SSB-SC *Acronym for SINGLE-SIDEBAND SUPPRESSED CARRIER.*

SSO *Acronym for SELF-SUSTAINED OSCILLATIONS.*

standard source A reference optical power source to which emitting and de-

---

Frequency Spectrum

FREQUENCY RANGE (Lower Limit Exclusive, Upper Limit Inclusive)		AMERICAN DESIGNATOR	ITU DESIGNATOR
Below 300 Hz	ELF	(Extremely Low Frequency)	—
300 – 3000 Hz	ILF	(Infra Low Frequency)	—
3 – 30 kHz	VLF	(Very Low Frequency)	4
30 – 300 kHz	LF	(Low Frequency)	5
300 – 3000 kHz	MF	(Medium Frequency)	6
3 – 30 MHz	HF	(High Frequency)	7
30 – 300 MHz	VHF	(Very High Frequency)	8
300 – 3000 MHz	UHF	(Ultra High Frequency)	9
3 – 30 GHz	SHF	(Super High Frequency)	10
30 – 300 GHz	EHF	(Extremely High Frequency)	11
300 – 3000 GHz	THF	(Tremendously High Frequency)	12

tecting devices may be compared for calibration purposes.

standing wave ratio (SWR) The ratio of the amplitude of a standing wave at an antinode to the amplitude at a node. *Note:* The standing wave ratio, SWR, in a uniform transmission line is:

$$SWR = (1 + RC)/(1 - RC)$$

where RC is the reflection coefficient. *See also: ANTI-NODE; NODE; REFLECTION COEFFICIENT; REFLECTION LOSS.*

star *See: D-STAR.*

star coupler A passive device in which power from one or several input waveguides is distributed amongst a larger number of output optical waveguides. *See also: OPTICAL COMBINER; TEE COUPLER.*

stark effect The splitting of spectral lines of electromagnetic radiation by an applied electric field.

state *See: EXCITED STATE; GROUND STATE.*

statistical multiplexing Multiplexing in which channels are established on a statistical basis. For example, connections are made according to probability of needs. *See also: CHANNEL.*

statistical time-division multiplexing Time-division multiplexing in which connections to communication circuits are made on a statistical basis. *See also: MULTIPLEX.*

step-index fiber A fiber in which there is an abrupt change in refractive index between the core and cladding along a fiber diameter, with the core refractive index higher than the cladding refractive index. *Note:* These may be more than one layer, each layer with a different refractive index that is uniform throughout the layer, with decreasing indices in the outside layer.

step index optical waveguide An optical waveguide having a step index profile. *See also: STEP INDEX PROFILE.*

step index profile A refractive index profile characterized by a uniform refractive index within the core and a sharp decrease in refractive index at the core-cladding interface. *Note:* This corresponds to a power-law profile with profile parameter, g, approaching infinity. *See also: CRITICAL ANGLE; DISPERSION; GRADED INDEX PROFILE; MODE VOLUME; MULTI-MODE OPTICAL WAVEGUIDE; NOR-MALIZED FREQUENCY; OPTICAL WAVEGUIDE; REFRACTIVE INDEX (OF A MEDIUM); TOTAL INTERNAL RE-FLECTION.*

stepwise variable optical attenuator A device that attenuates the intensity of lightwaves, when inserted into an optical waveguide link in discrete steps each of which is selectable by some means, such as by changing sets of cells, for example if fixed attenuation cells of 0, 3, 7, 17 dB are used three at a time, attenuations of 3, 6, 10, 13, 20, 23, and 27 dB attenuations are achievable.

steradian The unit solid angular measure, being the subtended surface area of a sphere divided by the square of the sphere radius. There are 4π steradians in a sphere. The solid angle subtended by a cone of half-angle A is 2π $(1-CosA)$ steradians.

stimulated Brillouin scattering When an incident beam is highly intense and monochromatic the Brillouin scattering process becomes stimulated. In amorphous materials stimulated Brillouin scattering (SBS) produces a backward wave. SBS limits the amount of power that can be carried in single-mode low-loss silica fibers to $\cong$5 mW. *Symbol: SBS. Note:* For examples see R. G. Smith, *Appl. Opt.,* vol. 11, pp. 2489–2494, Nov. 1972.

stimulated emission Radiation emitted when the internal energy of a quantum mechanical system drops from an excited level to a lower level when induced by the presence of radiant energy at the same frequency. An example is the radiation from an injection laser diode above lasing threshold. *See also: SPONTANE-OUS EMISSION.*

stimulated emission efficiency *See: INTERNAL QUANTUM EFFICIENCY. See: EXTERNAL DIFFERENTIAL QUAN-TUM EFFICIENCY.*

stimulated emission of radiation *See: MICROWAVE AMPLIFICATION BY STIMULATED EMISSION OF RADIA-TION.*

stimulated Raman scattering (SRS) When the incident beam is highly intense and monochromatic the Raman frequencies reach a threshold of amplification, and thus stimulated emission occurs. When using a monochromatic light source stimulated Raman scattering (SRS) in optical fibers occurs at one to two orders of magnitude more power than for stimulated Brillouin scattering.

145

Only for wide-spectrum light sources SRS limits the power handling capabilities of the fiber. Broad-band Raman outputs from Nd:YAG-pumped single-mode fibers are used for measuring pulse distortion in silica-based fibers, in the 1.06 to 1.6 μm wavelength range.

Note: For examples see L. G. Cohen and C. Lin, *IEEE J. Quantum Electron.*, vol. 14, pp. 855–858, Nov. 1979.

STL *Acronym for STANDARD TELE-GRAPH LEVEL.*

stop *See: APERTURE STOP; T-STOP.*

strength-member optical cable *See: CENTRAL STRENGTH-MEMBER OPTICAL CABLE; PERIPHERAL STRENGTH-MEMBER OPTICAL CABLE.*

stria A defect in optical materials, such as glass, plastic, or crystals, consisting of a more or less sharply defined streak of material having a slightly different index of refraction than the main body of the material.

Note: Striae usually cause wave-like distortions in objects seen through the material, exclusive of similar distortions due to variations in thickness or curvature. Striae are usually caused by temperature variation, or poor mixing of ingredients, causing the density (refractive index) to vary in different places.

strip-loaded diffused optical waveguide A three-dimensional optical waveguide, constructed from a two-dimensional diffused optical waveguide upon the surface of which has been deposited a dielectric strip of a lower refractive index material, thus confining the electromagnetic fields of the propagating mode to the vicinity of the strip hence achieving a three-dimensional guide.

stripe-geometry diode laser Diode laser for which the current is confined on the diode p-side (i.e., the one closest to the lasing region) to a metallic contact stripe defined by etching of an oxide film, proton bombardment or Zn diffusion through an n-type capping layer. Ranges of stripe widths for cw diode operation: 3-20 μm.

Note: The current may be confined by an internal stripe defined by back-biased current blocking layers.

stripper *See: CLADDING-MODE STRIPPER.*

stuffing *See: BIT STUFFING; DE-STUFFING.*

stuffing process *See: MOLECULAR STUFFING PROCESS (MS).*

submount A piece of material such as Si, BeO or diamond that is interposed between the diode laser and the copper heatsink (the mount) to relieve stresses induced by hard solders. The diode is soldered to the submount, and then the submount is soldered to the heatsink. *Synonym: INTERMEDIATE MOUNT.*

sudden ionospheric disturbance (SID) Abnormally high ionization densities in the D region caused by an occasional sudden outburst of ultraviolet light on the sun (solar flare). This results in a sudden increase in radio-wave absorption, which is most severe in the upper MF and lower HF frequencies. *See also: IONOSPHERE; IONOSPHERIC DISTURBANCE.*

supergroup *See: GROUP.*

superluminescent diode (SLD) A light-emitting diode (LED) with narrow spectral width, high-radiance, and fast response due to the fact that the diode has reached the threshold for amplification (i.e., onset of stimulated emission). An emitter based on stimulated emission with amplification but insufficient feedback for oscillation to build up.

Note: The SLD can serve as a source for optical fiber transmission systems. *See also: SPONTANEOUS EMISSION; STIMULATED EMISSION; THRESHOLD CURRENT DENSITY FOR AMPLIFICATION.*

superradiance Amplification of spontaneously emitted radiation in a gain medium, characterized by moderate line narrowing and moderate directionality.

Note: This process is generally distinguished from lasing action by the absence of positive feedback and hence the absence of well-defined modes of oscillation. *See also: LASER; SPONTANEOUS EMISSION; STIMULATED EMISSION.*

suppressed carrier transmission That method of communication in which the carrier frequency is suppressed either partially or to the maximum degree possible. One or both of the sidebands may be transmitted. *Synonym: REDUCED CARRIER TRANSMISSION. See also: DOUBLE-SIDEBAND SUPPRESSED CARRIER TRANSMISSION; SIDEBAND TRANSMISSION; TRANSMISSION.*

surface *See: OPTICAL SURFACE.*

surface-emitting LED A light-emitting diode for which the light is obtained from a surface parallel to the active layer. Surface-emitting LEDs have wider spectra and lower coupling efficiencies to optical fibers than edge-emitting LEDs.

Note: 1. For example, see D. Botez and G. Herskowitz, *Proc. IEEE*, vol. 68, pp. 689–732, June 1981.

Note: 2. Surface- and edge-emitting LEDs provide several milliwatts of power in the 0.8-1.2 micrometer spectral range at drive currents of 100–200 milliamperes; diode lasers at these currents provide tens of milliwatts. *Synonyms: FRONT-EMITTING LED; BURRUS LED. See also: EDGE-EMITTING LED.*

surface mirror *See: BACK-SURFACE MIRROR; FRONT-SURFACE MIRROR.*

surface refractivity The value of refractivity (refractive index) calculated from observations of pressure, temperature, and humidity at the earth's surface.

Note: The gradient of refractivity refers to the difference in refractivity over a given height, as between the surface and 100 meters, surface and 1 km, etc.

surface wave A wave that is guided by the interface between two different media or by a refractive index gradient in the medium. The field components of the wave may exist (in principle) throughout space (even to infinity) but become negligibly small within a finite distance from the interface.

Note: All guided modes, but not radiation modes, in an optical waveguide belong to a class known in electromagnetic theory as surface waves.

survey *See: PATH SURVEY.*

switched circuit A circuit that may be temporarily established at the request of one or more of the connected stations.

switched network A network providing switched communications service.

switched repetitively pulsed laser *See: Q-SWITCHED REPETITIVELY-PULSED LASER.*

switch-modulator *See: INTEGRATED-OPTICAL CIRCUIT FILTER-COUPLER-SWITCH-MODULATOR.*

SWR *Acronym for STANDING WAVE RATIO.*

symmetrical double-heterojunction diode *See: FOUR-HETEROJUNCTION DIODE.*

synchronization The process of adjusting corresponding significant instants of two signals to obtain a desired fixed relationship between these instants. *See also: AMPLITUDE QUANTIZED SYNCHRONIZATION; ANALOG SYNCHRONIZATION; BILATERAL SYNCHRONIZATION; DOUBLE-ENDED SYNCHRONIZATION; LINEAR ANALOG SYNCHRONIZATION; SINGLE-ENDED SYNCHRONIZATION; SYNCHRONOUS DATA-LINK CONTROL; SYNCHRONOUS DATA NETWORK; UNILATERAL SYNCHRONIZATION SYSTEM.*

synchronous network A network in which the clocks are controlled so as to run, ideally, at identical rates, or at the same mean rate with limited relative phase displacement.

Note: Ideally the clocks are synchronous, but they may be mesochronous in practice. By common usage such mesochronous networks are frequently described as synchronous. *See also: CLOCK; DATA TRANSMISSION.*

synchronous orbit An orbit in which a satellite has an orbital angular velocity synchronized with the rotational angular velocity of the earth and thus remains directly above a fixed point on the earth's surface.

Note: This occurs at an altitude of approximately 19,200 nautical miles over the equator. (1 nmi = 1.852 km) *See also: SATELLITE.*

synchronous satellite A satellite in a synchronous orbit.

synchronous system A system in which the transmitter and receiver are operating in a fixed time relationship. *See also: COMMUNICATIONS SYSTEM; HOMOCHRONOUS; MESOCHRONOUS; SYNCHRONOUS TRANSMISSION.*

synchronous TDM A generic type of multiplex in which timing is obtained from a clock, which in turn controls both the multiplexer and the channel source. *See also: ASYNCHRONOUS TIME-DIVISION MULTIPLEXING; TIME-DIVISION MULTIPLEX.*

synchronous transmission A transmission process such that between any two significant instants in the overall bit stream, there is always an integral number of unit intervals. *See also: ASYNCHRONOUS TRANSMISSION; SYNCHRONOUS SYSTEM.*

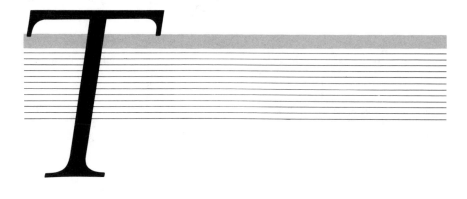

T$_0$ *See: THRESHOLD-CURRENT TEMPER-*
ATURE COEFFICIENT.

takeoff angle *Synonym: DEPARTURE*
ANGLE.

Talbot In the meter-kilogram-second
system of units, a unit of luminous energy equal to 10^7 lumergs and also equal to
1 lumen-second.

talk *See: CROSSTALK.*

tangential coupling The coupling of
one optical fiber to another by placing or
fusing the core of the fiber containing a
signal in close proximity for a short distance to another fiber core, to allow
some of the signal to leak or spill over to
the attached fiber, by subverting the
original signal-bearing fiber from keep-
ing all its light to itself.
Note: The degree of coupling is determined by the core-to-core spacing and
the fused length. This method of coupling also makes use of the evanescent
waves that are coupled to the waves in
the guide but are traveling on the outside of the optical waveguide. *Synonym:*
PICK-OFF COUPLING. See also: EVANES-
CENT FIELD COUPLING; LOOSE-TUBE
SPLICER.

tap A device for extracting a portion of
the optical signal from a fiber.

taper *See: OPTICAL TAPER.*

tapered fiber waveguide An optical
waveguide whose transverse dimensions
vary monotonically with length. *Syn-*
onym: TAPERED TRANSMISSION LINE.

tapered lens A lens whose cross section
shows a greater edge thickness on one
side than on the other.

tapered transmission line *Synonym:*
TAPERED FIBER WAVEGUIDE.

TDM *Acronym for TIME-DIVISION MUL-*
TIPLEX.

TDMA *Acronym for TIME-DIVISION*
MULTIPLE ACCESS.

TEA *Acronym for TRANSVERSE-EXCITED*
ATMOSPHERE LASER.

TEA laser *See: TRANSVERSE-ELECTRIC*
ATMOSPHERE LASER.

technique *See: COHERENT OPTICAL*
ADAPTIVE TECHNIQUE.

tee coupler A passive coupler that connects three ports. *See also: STAR COU-*
PLER.

telephoto lens An objective lens system
consisting of a positive and a negative
component separated from each other,
having such powers and separation that
the back focal length of the entire system is small in comparison with the
equivalent focal length.
Note: Such lenses are used for producing
large images of distant objects without
the necessity of a cumbersome length of
the instrument.

TEM mode *See: TRANSVERSE ELEC-*
TROMAGNETIC MODE.

TE mode *See: TRANSVERSE ELECTRIC*
MODE.

temperature *See: AMBIENT TEMPERA-*
TURE; ANTENNA GAIN-TO-NOISE
TEMPERATURE; ANTENNA NOISE
TEMPERATURE; COLOR TEMPERA-
TURE; EFFECTIVE INPUT NOISE TEM-
PERATURE; FRONT-END NOISE TEM-
PERATURE; KELVIN TEMPERATURE
SCALE; LUMINANCE TEMPERATURE;
RADIATION TEMPERATURE.

temperature coefficient *See: THRESH-*
OLD-CURRENT TEMPERATURE COEF-
FICIENT.

temporal coherence *See: COHERENT.*

temporally coherent light *See: TIME-*
COHERENT LIGHT.

temporally coherent radiation *See:*
COHERENT.

term *See: ATTENUATION TERM; PHASE*
TERM.

terminus *See: FIBER-OPTIC TERMINUS.*

terraced heterostructure large-opti-
cal-cavity (TH-LOC) diode laser
Mode-stabilized LOC-type diode laser
grown by one-step liquid-phase epitaxy

above misoriented channeled substrates of misorientation direction perpendicular to the channels' direction. The lasing cavity is formed by lateral thickness variations of the active and guide layers on the slope of a terrace. Typical reliable power levels: 15–25 mW/facet CW.

Note: For example see D. Botez and J. C. Connolly, *Appl. Phys. Lett.,* vol. 41, August 15th, 1982.

terraced-substrate (TS) diode laser Mode-stabilized DH diode laser fabricated by one-step liquid-phase epitaxy above a terrace etched into the substrate. The active layer has a local thickness increase on the terrace slope, thus forming a ridge waveguide in the lateral direction. Typical reliable power capability: 3–5 mW/facet CW.

Note: For examples see T. Sugino, *et al., IEEE J. Quantum Electron.,* vol. 15, August 1979.

TGM *Acronym for TRUNK GROUP MULTIPLEXER.*

theoretical resolving power The maximum possible resolving power determined by diffraction, frequently measured as an angular resolution determined from

$$A = 1.22 \left(\frac{B}{D} \right)$$

where A is the limiting resolution in radians, B is the wavelength of light at which the resolution is determined, and D is the diameter of the effective aperture.

theory *See: ELECTROMAGNETIC THEORY.*

thermal noise The noise generated by thermal agitation of electrons in a conductor. The noise power, P, is given by:

$$P = kT\Delta f$$

where P is in watts, k is Boltzmann's constant (1.3804×10^{-23} joules/kelvin), T is the conductor temperature in kelvins, and Δf is the bandwidth in hertz.

Note: Thermal noise power is distributed equally throughout the frequency spectrum. *Synonym: JOHNSON NOISE.*

thermal noise-limited operation The operation of a photodetector wherein the minimum detectable signal is limited by the thermal noise of the detector, the load resistance, and the amplifier noise.

thermal radiation The process of electromagnetic emission in which the radiated energy is extracted from the thermal excitation of atoms or molecules.

thermoplastic cement An adhesive whose viscosity decreases as the temperature is raised.

Note: Canada balsam, resin, and pitch are common thermoplastic optical cements.

thermosetting cement An adhesive that permanently sets or hardens at a certain high temperature.

Note: Methacrylate is a common thermosetting optical cement.

thick lens A lens whose axial thickness is so large that the principal points and the optical center cannot be considered as coinciding at a single point on its optical axis.

thin-film optical modulator A device made of multilayered films of material of different optical characteristics capable of modulating transmitted light by using electro-optic, electro-acoustic, or magneto-optic effects to obtain signal modulation.

Note: Thin-film optical modulators are used as component parts of integrated optical circuits.

thin-film optical multiplexer A multiplexer consisting of layered optical materials that make use of electro-optic, electro-acoustic, or magneto-optic effects to accomplish the multiplexing.

Note: Thin-film optical multiplexers may be component parts of integrated optical circuits.

thin-film optical switch A switching device for performing logic operations using light waves in thin films, usually supporting only one propagation mode, making use of electro-optic, electro-acoustic, or magneto-optic effects to perform switching functions, such as are performed by semiconductor gates (AND, OR, NEGATION).

Note: Thin-film optical switches may be component parts of integrated optical circuits. *See: ELECTRO-OPTIC DIRECTIONAL COUPLER (EDC) SWITCH.*

thin-film optical waveguide An optical waveguide consisting of thin layers of differing refractive indices, the lower indexed material on the outside or as a substrate, for supporting usually a single electromagnetic wave propagation mode.

Note: The thin-film waveguide lasers, modulators, switches, directional couplers, filters, and related components need to be coupled from their integrated optical circuits to the optical waveguide transmission media, such as optical fibers and slab dielectric waveguides. *See also: OPTICAL WAVEGUIDE.*

thin-film waveguide A transparent dielectric film, bounded by lower index materials, capable of guiding light. *See also: OPTICAL WAVEGUIDE.*

thin lens A lens whose axial thickness is sufficiently small so that the principal points, the optical center, and the vertices of its two surfaces can be considered as coinciding at the same point on its optical axis.

TH-LOC laser *See: TERRACED HETERO-STRUCTURE LARGE-OPTICAL-CAVITY DIODE LASER.*

threshold current The driving current corresponding to lasing threshold. *See also: LASING THRESHOLD.*

threshold current density for amplification In a diode, level of injected current per unit area of active region at which the net electronic gain is zero (i.e., population inversion is realized). At the threshold current density for amplification the diode changes from the LED mode to the superradiant mode (i.e., superluminescent (SLD) diode).
Note: In diode lasers the threshold for amplification occurs within 30% of the lasing threshold. *See: SUPERLUMINESCENT (SLD) DIODE.*

threshold current density for lasing In a diode laser, level of injected current per unit area of active region at which the net electronic gain equals the cavity round-trip losses (i.e., internal cavity losses + mirror losses). Expressed in kA/cm^2. Symbol: J_{th}. *See also: LASING THRESHOLD.*

threshold-current temperature coefficient Coefficient in the empirical exponential formula describing the threshold-current temperature dependence over a given temperature interval: $I_{th}(T) = I_{th}(0°C) \cdot \exp(T/T_0)$, where T is the temperature in degrees Celsius, and T_0 is the threshold-current temperature coefficient. T_0 has the same value in degrees Celsius and in kelvin. *Synonym: CHARACTERISTIC TEMPERATURE.*
Note: For AlGaAs and InGaAsP DH lasers typical T_0 values are 120–180°C and 40–70°C, respectively, over the 20 to 70°C temperature interval.
Note: T_0 is usually defined for the pulsed threshold current.

time coherence *See: COHERENT.*

time-coherent light Light that has the property that at any point in time, i.e., any instant, the amplitude, phase, and time variation are predictable, the prediction being based on the amplitude, phase, and time variation at a previous time. *Synonym: TEMPORALLY COHERENT LIGHT. See also: COHERENT-LIGHT; SPACE-COHERENT LIGHT.*

time-delay distortion *Synonym: DELAY DISTORTION.*

time-division multiple access (TDMA) In satellite communications, the use of time interlacing to provide multiple and apparently simultaneous transmissions to a single transponder with a minimum of interference.

time-division multiplex (TDM) A method of deriving two or more apparently simultaneous channels from a given frequency spectrum of a transmission medium connecting two points by assigning discrete time intervals in sequence to each of the individual channels. During a given time interval the entire available frequency spectrum can be used by the channel to which it is assigned.
Note: In general, time division multiplex systems use pulse transmission. The multiplex pulse train may be considered to be the interleaved pulse trains of the individual channels. The individual channel pulses may be modulated either in an analog or a digital manner. *See also: ASYNCHRONOUS TIME-DIVISION MULTIPLEXING (ATDM); MULTIPLEX AGGREGATE BIT RATE; SYNCHRONOUS TDM.*

time division switching A switching arrangement whereby the various connections share a common path, but are separated in time. This is accomplished by permitting each connection to use the common path in sequence for a short period of time. *See also: SWITCHING; SYSTEM.*

time-domain reflectometer Optical instrument used for OTDR, composed of a laser, an oscilloscope, and driving electronics.

time-domain reflectometry *See: OTDR.*

TJS laser *See: TRANSVERSE-JUNCTION-STRIPE DIODE LASER.*

TL *Acronym for TRANSMISSION LEVEL.*

TM mode *Abbreviation for TRANSVERSE MAGNETIC MODE.*

T-number The equivalent F-number of a fictitious lens that has a circular opening and 100 percent transmittance, and that gives the same central illumination as the actual lens under consideration. Mathematically,

$$\text{T-Number} = \frac{\text{EFL}}{\text{Dia of T-stop}}$$

$$\text{or T-Number} = \frac{\text{EFL}}{2}\left(\frac{\pi}{\text{AT}}\right)^{1/2}$$

where EFL is the equivalent focal length, A is the area of the entrance pupil, T is the transmittance of the lens system, and $\pi = 3.1416$.

tolerance field 1. In general, the region between two curves (frequently two circles) used to specify the tolerance on component size. 2. When used to specify fiber cladding size, the annular region between the two concentric circles of diameter $D + \Delta D$ and $D - \Delta D$. The first circumscribes the outer surface of the homogeneous cladding; the second (smaller) circle is the largest circle that fits within the outer surface of the homogeneous cladding. 3. When used to specify the core size, the annular region between the two concentric circles of diameter $d + \Delta d$ and $d - \Delta d$. The first circumscribes the core area; the second (smaller) circle is the largest circle that fits within the core area. *Note:* The circles of definition 2 need not be concentric with the circles of definition 3. *See also: CLADDING; CORE; CONCENTRICITY ERROR; HOMOGENEOUS CLADDING.*

total channel noise The sum of random noise and intermodulation noise plus crosstalk. *Note:* Impulse noise is not included because different techniques are required for its measurement.

total diffuse reflectance *See: DIFFUSE REFLECTANCE.*

total harmonic distortion The ratio of the sum of the powers of all harmonic frequency signals (other than the fundamental) to the power of the fundamental frequency signal. This ratio is measured at the output of a device under specified conditions and is expressed in dB. *See also: HARMONIC DISTORTION; SINGLE HARMONIC DISTORTION.*

total internal reflection The total reflection that occurs when light strikes an interface at angles of incidence (with respect to the normal) greater than the critical angle. *See also: CRITICAL ANGLE; STEP INDEX OPTICAL WAVEGUIDE.*

total internal reflection angle *See: CRITICAL ANGLE.*

total radiation temperature The temperature at which a blackbody radiates a total amount of electromagnetic radiation flux equal to that radiated by the body whose total radiation temperature is being considered.

transducer A device that converts one form of energy (optical, electrical, thermal, or mechanical) into another. Used for thin-film devices.

transfer characteristics Those intrinsic parameters of a system, subsystem, or equipment which, when applied to the input of the system, subsystem, or equipment, will fully describe its output.

transfer function (of a device) The complex function, H(f), equal to the ratio of the output to input of the device as a function of frequency. The amplitude and phase responses are, respectively, the magnitude of H(f) and the phase of H(f). *Note: 1.* For an optical fiber, H(f) is taken to be the ratio of output optical *power* to input optical *power* as a function of modulation frequency. *Note: 2.* For a linear system, the transfer function and the impulse response h(t) are related through the Fourier transform pair, a common form of which is given by

$$H(f) = \int_{-\infty}^{\infty} h(t)\exp(i2\pi ft)\,dt$$

and

$$h(t) = \int_{-\infty}^{\infty} H(f)\exp(-2\pi ft)\,df$$

where f is frequency. Often H(f) is normalized to H(0) and h(t) to

$$\int_{-\infty}^{\infty} h(t)dt, \text{ which by definition is H(0)}.$$

Synonyms: BASEBAND RESPONSE FUNCTION; FREQUENCY RESPONSE. See also: IMPULSE RESPONSE.

transfer rate *See: DATA TRANSFER RATE.*

transimpedance *See: OPTICAL TRANSIMPEDANCE.*

transition frequency The frequency associated with two discrete energy levels in an atomic system. *Note:* The transition frequency associated with energy levels E_2 and E_1. E_2 greater than E_1 is

$$\nu\,(2,1) = \frac{(E_2 - E_1)}{\nu}$$

where E_2 and E_1 are the energy levels, h is Planck's Constant, and ν is the frequency associated with the two levels. If a transition from E_2 to E_1 occurs, a photon is likely to be emitted.

transition zone The zone between the end of the near field region of an antenna and the beginning of the far field region.

transmission loss Total loss encountered in transmission through a system. *See also: ATTENUATION; OPTICAL DENSITY; REFLECTION; TRANSMITTANCE.*

transmission system *See: FIBER-OPTIC TRANSMISSION SYSTEM; LASER FIBER-OPTIC TRANSMISSION SYSTEM.*

transmission window *Synonym: SPECTRAL WINDOW.*

transmissivity The transmittance of a unit length of material, at a given wavelength, excluding the reflectance of the surfaces of the material; the intrinsic transmittance of the material, irrespective of other parameters such as the reflectances of the surfaces. No longer in common use. *See also: TRANSMITTANCE.*

transmittance The ratio of transmitted power to incident power.
Note: In optics, frequently expressed as optical density or percent; in communications applications, generally expressed in dB. Formerly called "transmission." *See also: ANTIREFLECTION COATING; OPTICAL DENSITY; TRANSMISSION LOSS.*

transmittancy The ratio of the transmittance of a solution to that of an equal thickness of the solvent alone.

transmitting element The radiating terminus at an optical junction.

transverse electric (TE) mode A mode whose electric field vector is normal to the direction of propagation.
Note: In an optical fiber, TE and TM modes correspond to meridional rays. *See also: MERIDIONAL RAY; MODE.*

transverse electromagnetic (TEM) mode A mode whose electric and magnetic field vectors are both normal to the direction of propagation. *See also: MODE.*

transverse-excited atmosphere laser

(TEA) A carbon-dioxide or other gas laser in which the electric field excitation of the active medium is transverse (across) to the flow of the active medium.
Note: This type of laser operates in a gas pressure range higher than that required for longitudinal excitation.

transverse interferometry The method used to measure the index profile of an optical fiber by placing it in an interferometer and illuminating the fiber transversely to its axis. Generally, a computer is required to interpret the interference pattern. *See also: INTERFEROMETER.*

transverse-junction-stripe (TJS) diode laser Mode-stabilized diode laser fabricated by controlled deep Zn diffusions in planar LPE-grown material. The junctions across which carriers are injected are perpendicular to the grown layers. Lateral mode control is provided by two Zn-diffusion fronts. Typical reliable power levels: 1 to 3 mW/facet CW.
Note: For examples see W. Susaki, *et al.*, *IEEE J. Quantum Electron.*, vol. 13, pp. 587–591, Aug. 1977.

transverse magnetic (TM) mode A mode whose magnetic field vector is normal to the direction of propagation.
Note: In a planar dielectric waveguide (as within an injection laser diode), the field direction is parallel to the core-cladding interface. In an optical waveguide, TE and TM modes correspond to meridional rays. *See also: MERIDIONAL RAY; MODE.*

transverse mode control In diode lasers, fabrication procedures via which stable operation in a single transverse mode, usually the fundamental one, is obtained. For DH lasers fundamental mode control is simply achieved by growing structures of thin active layers (i.e., $D \leq TL$). In LOC structures transverse mode control is a function of active- and guide-layer thicknesses, lateral geometry, and refractive-index step between the active and guide layers.
Note: In some references used to define lateral mode control.

transverse mode(s) In diode lasers spatial modes in a direction perpendicular to the active layer. Most DH and SH lasers operate in the fundamental transverse mode. LOC-type lasers can operate in high-order transverse modes.
Note: In some references term used to

define spatial modes in the lateral direction.

transverse offset loss   *Synonym: LATERAL OFFSET LOSS.*

transverse propagation constant   The propagation constant evaluated along a direction perpendicular to the waveguide axis.
Note: The transverse propagation constant for a given mode can vary with the transverse coordinates. *See also: PROPAGATION CONSTANT.*

transverse scattering   The method for measuring the index profile of an optical fiber or preform by illuminating the fiber or preform coherently and transversely to its axis, and examining the far-field irradiance pattern. A computer is required to interpret the pattern of the scattered light. *See also: SCATTERING.*

trap   *See: OPTICAL FIBER TRAP.*

trapped mode   *See: BOUND MODE.*

trapped ray   *Synonym: GUIDED RAY.*

triple mirror   Three reflecting surfaces, mutually at right angles to each other, arranged like the inside corner of a cube. *Note:* The triple mirror may be constructed of solid glass in which case the transmitting face is normal to the diagonal of the cube, or it may consist of the three plane mirrors supported in a precisely constructed metal framework. The triple reflector has a constant deviation of 180° for all angles of incidence, hence a ray of light incident from any angle is reflected back parallel to itself. *Synonym: CORNER-CUBE REFLECTOR; CORNER REFLECTOR; RETRODIRECTIVE REFLECTOR.*

troposcatter   *Synonym: TROPOSPHERIC SCATTER.*

troposphere   The layer of the earth's atmosphere, between the surface and the stratosphere, in which about 80 percent of the total mass of air is concentrated and in which temperature normally decreases with altitude.
Note: The thickness of the troposphere varies with season and latitude; it is usually 16 km to 18 km over tropical regions and 10 km or less over the poles.

tropospheric scatter   1. The propagation of radio waves by scattering as a result of irregularities or discontinuities in the physical properties of the troposphere. 2. A method of transhorizon communications utilizing frequencies from approximately 350 MHz to approximately 8400 MHz.

Note: The propagation mechanism is still not fully understood, though it includes several distinguishable but changeable mechanisms such as propagation by means of random reflections and scattering from irregularities in the dielectric gradient density of the troposphere, smooth-earth diffraction, and diffraction over isolated obstacles (knife-edge diffraction). *Synonym: TROPOSCATTER.*

tropospheric wave   A radio wave that is propagated by reflection from a place of abrupt change in the dielectric constant or its gradient in the troposphere.
Note: In some cases the ground wave may be so altered that new components appear to arise from reflection in regions of rapidly changing dielectric constant. When these components are distinguishable from the other components, they are called tropospheric waves.

true field   *See: VIEW FIELD.*

trunk   A single transmission channel between two points, both of which are switching centers or nodes, or both. *See also: TRANSMISSION CHANNEL.*

trunk group   Two or more trunks of the same type between the same two points. *See also: GROUP.*

trunk group multiplexer (TGM)   A time division multiplexer whose function is to combine individual digital trunk groups into a higher rate bit stream for transmission over wideband digital communication links. *See also: MULTIPLEX.*

TS laser   *See: TERRACED-SUBSTRATE DIODE LASER.*

T-stop   The equivalent, perfectly transmitting, circular opening of diameter D such that

$$\pi \left(\frac{D}{2}\right)^2 = TA$$

where A is the area of the entrance pupil of the objective, T is the transmittance of the lens system, and $\pi = 3.1416$.

tube photometer   *See: PHOTOEMISSIVE TUBE PHOTOMETER.*

tube splicer   *See: LOOSE-TUBE SPLICER.*

tunable laser   An organic dye or parametric-oscillator laser whose emission can be varied continuously across a broad spectral range.
Note: Sometimes applied to carbon-dioxide or other molecular lasers whose emission can be tuned to one of several wavelengths (spectral lines).

tunneling mode *Synonym:* *LEAKY MODE.*

tunneling ray *Synonym: LEAKY RAY.*

turn-on delay In diode lasers, delay in the light output response with respect to the applied current pulse. Turn-on delay is the time it takes the injected carriers to reach the carrier density needed for lasing threshold. Typical values: 2–8 ns, and decreasing with increasing current above threshold. For high-speed (e.g., 0.2–2 GHz), diode-laser modulation turn-on delay is eliminated by dc biasing the device at or near threshold.
Note: For examples see S. M. Sze, *Physics of Semiconductor Devices,* 2nd edition, Wiley & Sons, New York, 1982. *Synonym:*
"SWITCH-ON" DELAY.

two-way alternate operation *Synonym: HALF-DUPLEX OPERATION.*

two-way simultaneous operation *Synonym: DUPLEX OPERATION.*

two-wire circuit A circuit formed by two metallic conductors insulated from each other.
Note: The term is also used, in contrast to a four-wire circuit, to indicate a circuit using one line or channel for communications in both directions.

Twyman-Green interferometer A testing device in which the observer sees a contour map of the emergent wavefront in terms of the wavelength of the light used in the task.

ultraviolet absorption edge Intrinsic absorption in silica-based fibers that decreases with increasing wavelength. Referred to as ultraviolet (UV) edge. In silica-based fibers at wavelengths above 1 μm the UV absorption edge is negligible by comparison to Rayleigh scattering. Of importance only in the visible.
Note: For examples see D. Botez and G. Herskowitz, *Proc. IEEE,* vol. 68, pp. 689–732, June 1980.

ultraviolet fiber optics Fiber optics involving the use of ultraviolet (UV) light-conducting components designed to transmit electromagnetic waves shorter in wavelength than the waves in the visible region of the spectrum.
Note: Primary applications include medical technology, medicine, physics, materials testing, photochemistry, genetics, and many other fields. Optical fibers with high UV transmittance have been developed and are being used.

ultraviolet light Rays of electromagnetic radiant energy immediately beyond the violet end of the visible spectrum and in of the order of 390 to 100 nanometers in wavelength.

ultraviolet light guide Special optical materials in various geometric shapes, such as tubes, cylinders, sheets, and fibers that have the special capability of transmitting light in the ultraviolet (UV) region of the spectrum, i.e., with a wavelength of the order of 200 to 300 nanometers which is less than the wavelength of the visible spectrum, used in fiber optics, about 0.9 to 1.0 micrometers.
Note: UV light guides are primarily used in medicine, biochemistry, microscopy, physiology, and medical engineering.

ultraviolet (UV) The region of the electromagnetic spectrum between the short wavelength extreme of the visible spectrum (about 0.4 μm) and 0.04 μm. *See also: ULTRAVIOLET; LIGHT.*

unbalanced line A transmission line in which the magnitudes of the voltages on the two conductors are not equal with respect to ground; for example, a coaxial line.

unbalanced modulator A modulator in which the modulation factor is different for the alternate half-cycles of the carrier. *Synonym: ASYMMETRICAL MODULATOR. See also: MODULATION FACTOR.*

unbound mode Any mode that is not a bound mode; a leaky or radiation mode of the waveguide. *Synonym: RADIATIVE MODE. See also: BOUND MODE; CLADDING MODE; LEAKY MODE.*

unidirectional channel *Synonym: ONE-WAY-ONLY CHANNEL.*

unidirectional operation A method of operation in one direction only between terminals, one of which is a transmitter and the other a receiver.

uniform density lens A layered lens or blank, one layer of which is clear, and the other of absorptive-type glass, the clear portion being surfaced to the desired curvature, while the thickness of the tinted layer remaining constant, which results in a lens with the same shade, i.e. transmittance, in the center as in the periphery.

uniform encoding An analog-to-digital conversion process in which all of the quantization subrange values are equal. *Synonym: UNIFORM QUANTIZING. See also: ANALOG ENCODING; QUANTIZATION; QUANTIZATION LEVEL; SIGNAL.*

uniform-index profile In materials used for optical transmission, such as an optical fiber, a uniform linearly decreasing index of refraction from the inside radially toward the outside.

uniform-index-profile fiber A graded index optical fiber in which the refractive index varies linearly from the center of the fiber radially to the outside surface, with a lower index at the outside surface. *See also: GRADED INDEX FIBER.*

uniform Lambertian   A lambertian distribution that is uniform across a specified surface.

uniform quantizing   *Synonym: UNIFORM ENCODING.*

unintelligible crosstalk   Crosstalk giving rise to unintelligible sounds.

uplink   That portion of a communications link used for transmission of signals from an earth terminal to a satellite or airborne platform. It is the converse of DOWNLINK. *See also: Link; SATELLITE.*

Urbach edge   *See: UV EDGE.*

UV   *Acronym for ULTRAVIOLET.*

UV edge   *See: ULTRAVIOLET ABSORPTION EDGE.*

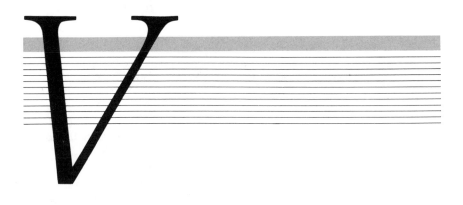

VAD *Acronym for VAPOR PHASE AXIAL DEPOSITION.*

valence band A filled or virtually filled energy band in the electronic-energy band scheme characterizing a solid. All valence bands are below the conduction band. Electrons are usually excited from the topmost valence band into the conduction band, leaving behind holes. *See: CONDUCTION BAND; HOLE.*

vapor deposition process *See: MODIFIED CHEMICAL VAPOR-DEPOSITION PROCESS; PLASMA-ACTIVATED CHEMICAL-VAPOR DEPOSITION PROCESS.*

vapor-phase epitaxy (VPE) Single-crystal growth technique during which atomic species are carried in the gas phase and deposited onto a single-crystal substrate. The deposited material is a product of chemical reactions. Two major types of VPE are used: metal chloride and organometallic. Both are used for the fabrication of light sources and photo-detectors.

vapor-phase oxidation process *See: AXIAL VAPOR PHASE OXIDATION PROCESS; CHEMICAL VAPOR-PHASE OXIDATION PROCESS; INSIDE VAPOR-PHASE OXIDATION PROCESS; MODIFIED INSIDE VAPOR-PHASE OXIDATION PROCESS; OUTSIDE VAPOR-PHASE OXIDATION PROCESS.*

variable optical attenuator *See: CONTINUOUS VARIABLE OPTICAL ATTENUATOR; STEPWISE VARIABLE OPTICAL ATTENUATOR.*

vector *See: ELECTRIC VECTOR.*

Vee value *See: ABBE CONSTANT.*

velocity *See: GROUP VELOCITY; PHASE VELOCITY.*

velocity of light This term usually refers to the speed of monochromatic light waves, i.e., to the phase velocity.

Note: The velocity of light in vacuum is

299,792.5 kilometers per second. The phase velocity in a medium is

$$\frac{C_0}{N}$$

where N is the refractive index of the medium and C_0 is the velocity of light (in a vacuum) given above.

vestigial sideband transmission A modified double-sideband transmission in which one sideband, the carrier, and only a portion of the other sideband are transmitted.

V-groove diode laser Mode-stabilized diode laser for which tight current confinement and/or lateral mode control is realized via a Zn-diffusion into a V-shaped channel, etched from the top of a standard planar DH structure. The proximity of the Zn-diffusion front determines if the optical mode is fully gain-guided (as in a narrow-stripe laser) or partially gain-guided and partially index-antiguided. Lateral far-fields are astigmatic and two-peaked. Typical reliable output powers: 5–10 mW/facet cw.

Note: For example see C. Wolk, *et al., IEEE J. Quantum Electron.,* vol. QE-17, pp. 756–759, May 1981.

V-grooved substrate buried heterostructure diode laser Mode-stabilized InGaAsP/InP diode laser grown by one-step liquid-phase epitaxy above a V-groove originally etched into the substrate. The active layer is crescent-shaped and provides positive-index confinement. Typical reliable power levels: 3–5 mW/facet CW. *See also: BURIED-CRESCENT DIODE LASER.*

Note: For example see H. Ishikawa, *et al., Electron. Lett.,* vol. 17, pp. 465–466, June 25th, 1981.

video disc *See: OPTICAL VIDEO DISC.*

view In satellite communications, the

ability of a satellite station to "see" a satellite, having it sufficiently above the horizon and clear of other obstructions so that it is within a free line of sight from the satellite station.

Note: A pair of satellite stations has a satellite in "mutual" view when both have unobstructed line-of-sight contact with the satellite simultaneously. *See also: SATELLITE.*

view field In general, the maximum cone or fan of rays passed through an aperture and measured at a given vertex.

Note: In an instrument the field of view is synonymous with true field.

vignetting The loss of light through an optical element due to the entire bundle of light rays not passing through.

virtual image The point from which a bundle of divergent light rays appear to proceed when the rays have a given divergence but no real physical point of intersection.

Note: The distance of the virtual image is inversely proportional to the divergence of the rays. Since there is no physical intersection of rays there is no real image that can be focused on a screen. The image of any real object produced by a negative lens or convex mirror is always virtual. The image produced by a positive lens of an object located within its focal length is also virtual.

visible spectrum *See: LIGHT.*

visual spectrum The band of color produced by decomposing white light into its components by the process of dispersion.

Note: The rainbow is an example of a spectrum produced by the dispersion of white light by water droplets. *See also: ELECTROMAGNETIC SPECTRUM.*

vitreous silica Glass consisting of almost pure silicon dioxide (SiO_2). *Synonym: FUSED SILICA. See also: FUSED QUARTZ.*

V number *Synonym: NORMALIZED FREQUENCY.*

voice-plus circuit *Synonym: COMPOSITED CIRCUIT.*

voltage standing wave ratio (VSWR) The ratio of maximum to minimum voltage in the standing wave pattern that appears along a transmission line. It is used as a measure of impedance mismatch between the transmission line and its load.

volume *See: MODE VOLUME.*

volume unit (VU) The unit of measurement for electrical speech power in communication work as measured by a vu meter in the prescribed manner.

Note: The VU meter is a volume indicator in accordance with American National Standard C16.5–1942. It has a scale and specified dynamic and other characteristics in order to obtain correlated readings of speech power necessitated by the rapid fluctuation in level of voice signals. Zero vu equals zero dBm (1 milliwatt) in measurements of sine wave test tone power.

VPE *Acronym for VAPOR-PHASE EPITAXY.*

VSB laser *See: V-GROOVED SUBSTRATE BURIED HETEROSTRUCTURE DIODE LASER.*

VSWR *Acronym for VOLTAGE STANDING WAVE RATIO.*

VU *Acronym for VOLUME UNIT.*

watch *See: LENS WATCH.*

wavefront The locus of points having the same phase at the same time.

wavefront control The performing of operations in an optical system so as to manipulate the shape of the wavefront of an electromagnetic wave, usually in the visible and near visible region of the frequency spectrum, and usually with the intent of obtaining clear images of illuminated objects, i.e., of obtaining a spherical wave front.
Note: Among the methods of wavefront control are phase conjugation, aperture tagging, wavefront compensation, and image sharpening.

waveguide 1. Any structure capable of confining the energy of an electromagnetic wave to a specific relatively narrow controllable path capable of being varied or altered, such as a metal pipe of rectangular cross section, an optical glass fiber of circular cross section, a thin film of semiconductor material, or a coaxial cable. 2. A transmission line consisting of a hollow metallic conductor, generally rectangular, elliptical, or circular in shape, within which electromagnetic waves may be propagated.
Note: The guide may, under certain conditions, be made with a solid dielectric or a gas-filled dielectric conductor. *See also: CLOSED WAVEGUIDE; MULTIMODE WAVEGUIDE; OPEN WAVEGUIDE; DIFFUSED OPTICAL WAVEGUIDE; STRIP-LOADED DIFFUSED OPTICAL WAVEGUIDE; FIBER OPTICAL WAVEGUIDE; HETEROEPITAXIAL OPTICAL WAVEGUIDE; SLAB-DIELECTRIC OPTICAL WAVEGUIDE; THIN-FILM OPTICAL WAVEGUIDE.*

waveguide delay distortion In an optical waveguide, such as an optical fiber, dielectric slab waveguide, or an integrated optical circuit, the distortion in received signal caused by the differences in propagation time for each wavelength, i.e. the delay versus wavelength effect for each propagating mode, causing a spreading of the total received signal at the detector.
Note: Waveguide delay distortion contributes to group-delay distortion, along with material dispersion and multimode group-delay spread.

waveguide dispersion For each mode in an optical waveguide, the term used to describe the process by which an electromagnetic signal is distorted by virtue of the dependence of the phase and group velocities on wavelength as a consequence of the geometric properties of the waveguide. In particular, for circular waveguides, the dependence is on the ratio (a/λ), where a is core radius and λ is wavelength. *See also: DISPERSION; DISTORTION; MATERIAL DISPERSION; MULTIMODE DISTORTION; PROFILE DISPERSION.*

waveguide mode *See: DEGENERATE WAVEGUIDE MODE.*

waveguide scattering Scattering (other than material scattering) that is attributable to variations of geometry and index profile of the waveguide. *See also: MATERIAL SCATTERING; NONLINEAR SCATTERING; RAYLEIGH SCATTERING; SCATTERING.*

wavelength 1. The length of a wave measured from any point on one wave to the corresponding point on the next wave; such as from crest to crest.
Note: Wavelength determines the nature of the various forms of radiant energy that comprise the electromagnetic spectrum; for example, it determines the color of light.
2. For a sinusoidal wave, the distance between points of corresponding phase of two consecutive cycles.

Note: the wavelength λ is related to the phase velocity v and the frequency f by

$$\lambda = \frac{v}{f}$$

See also: LIGHT; PEAK WAVELENGTH.

wavelength division multiplexing (WDM) The provision of two or more channels over a common optical waveguide, the channels being differentiated by optical wavelength.

wave number The reciprocal of wavelength, i.e., the number of wavelengths per unit distance in the direction of propagation of a wave. Sometimes defined as the reciprocal of wavelength multiplied by 2π.

wave object *See: SINE WAVE OBJECT.*

WDM *Acronym for WAVELENGTH DIVISION MULTIPLEX.*

weakly guiding fiber A fiber for which the difference between the maximum and the minimum refractive index is small (usually less than 1%).

Weibull distribution Statistical distribution used for characterizing device failures in time or fiber failure as a function of strain. The cumulative distribution function is $F(t) = 1 - \exp(-\alpha t^m)$, where m is the distribution shape parameter, and t is time or stress. Cumulative percent failures are plotted on special graph paper, referred to as Weibull paper, for which one axis is proportional to $\ln t$ and the other is proportional to $\ln \ln [1/(1 - F(t))]$. The slope of a curve plotted on Weibull paper gives the distribution shape parameter, m.
Note: For examples see H. Imai, *et al., J. Appl. Phys.,* vol. 52, pp. 3167–3171, May 1981.

weighting *See: C-MESSAGE WEIGHTING; FLAT WEIGHTING; F1A-LINE WEIGHTING; HA1-RECEIVER WEIGHTING; NOISE WEIGHTING; PSOPHOMETRIC WEIGHTING; WEIGHTING NETWORK; 144-LINE WEIGHTING; 144-RECEIVER WEIGHTING (Numeric entries are indexed after alphabetic entries in this document).*

W-guide Dielectric waveguide for which the refractive index profile has a local maximum placed between two minima. The waveguiding structure for W-type fibers or for the lateral wave confinement in certain types of mode-stabilized diode lasers (e.g., CDH-LOC diode laser). In a lasing structure a W-guide allows large lateral spot size while being strongly discriminatory against high-order-mode oscillation.

white light Electromagnetic radiation having a spectral energy distribution that produces the same color sensation to the average human eye as average noon sunlight.

White noise Noise whose frequency spectrum is continuous and uniform over a wide frequency range. *Synonym: ADDITIVE WHITE GAUSSIAN NOISE.*

wideband modem 1. A modem whose modulated output signal can have an essential frequency spectrum that is broader than that which can be wholly contained within, and faithfully transmitted through, a voice channel with a nominal 4 kHz bandwidth. 2. A modem whose bandwidth capability is greater than that of a narrowband modem. *See also: MODEM; NARROWBAND MODEM.*

wideband system A system with a multichannel bandwidth of 4 kHz or more. *Synonym: BROADBAND SYSTEM. See also: CHANNEL BANK; CHANNELIZATION; COMMUNICATIONS SYSTEM; GROUP.*

width *See: LASER PULSE LENGTH.*

window *See: SPECTRAL WINDOW.*

"window-stripe" deep-Zn diffused diode laser Deep-Zn diffused diode laser of nonabsorbing-mirror (NAM) type. The nonabsorbing-mirror regions are obtained by preferential Zn diffusion in the longitudinal direction. The diode's mirrors are formed by cleaving in the nonabsorbing regions. *See: NAM LASER; CRANK-TYPE TJS DIODE LASER.*
Note: For examples see M. Ueno, *IEEE J. Quantum Electron.,* vol. QE-17, pp. 2113–2122, October 1981.

wire *See: CHICKEN WIRE.*

worst hour of the year That hour of the year during which the median noise over any radio path is at a maximum.
Note: This hour is considered to coincide with the hour during which the greatest transmission loss occurs.

W-type fiber Fiber having two claddings of which the inner one is of lower refractive index than the outer one. Used both for multimode and single-mode fibers to provide both discrimination against high-order modes as well as negative pulse dispersion. *Synonym: DEPRESSED CLADDING (DC) FIBER.*
Note: For examples see Onoda, *et al., Opt. Comm.,* vol. 17, pp. 201–203, May 1976.

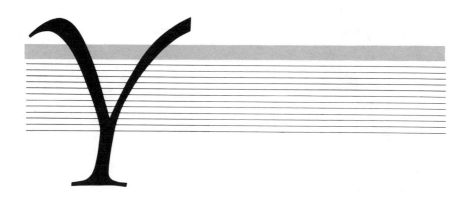

YAG/LED source A laser light source used for optical fiber transmission consisting of a Neodymium (Nd) Yttrium Aluminum Garnet (YAG) crystal laser usually pumped by a Light-Emitting Diode (LED).

Note: A YAG/LED source emits in a stable, single mode in frequency. However, the source is inefficient and bulky.

Y coupler *See: TEE COUPLER.*

yttrium aluminum garnet source *See: YAG/LED SOURCE.*

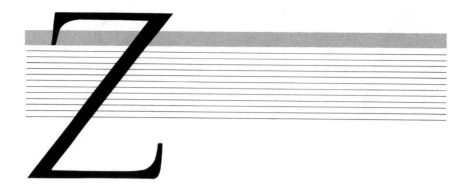

Zeeman effect The splitting of electromagnetic radiation into its component frequencies, i.e., the splitting of spectral wavelengths (lines) by an applied magnetic field.

zone *See: COMMUNICATIONS ZONE; FRESNEL ZONE; SKIP ZONE; ZONE OF SILENCE.*

zone of silence *Synonym: SKIP ZONE.*

zoom lens An optical system that has components that move in such a way as to change the focal length, while maintaining a fixed image position, thus the image size can be varied while leaving the optical system in a fixed position. *Synonym: PANCRATIC LENS.*

ACRONYMS
AND ABBREVIATIONS

Å	Angstrom
ACVD	Activated chemical vapor deposition process
A–D	Analog-to-digital
AM	Amplitude modulation
AME	Amplitude modulation equivalent
AMI	Alternate mark inversion
AP	Anomalous propagation
APC	Adaptive predictive coding
APD	Avalanche photodiode
ASCII	American national standard code for information interchange
AVPO	Axial vapor-phase oxidation process
AWGN	Additive white Gaussian noise
BC	Buried-crescent diode laser
BCI	Bit count integrity
Bd	Baud
BER	Bit error rate
BFL	Back focal length
BH	Buried-heterostructure diode laser
BH-LOC	Buried-heterostructure large-optical-cavity diode laser
bit	Binary digit
bit/s	Bits per second
CCH	Connections per circuit hour
CCIR	International radio consultative committee
CCM	Close-confinement-mesa diode laser
CCS	Hundred-call-seconds
CDF	Combined distribution frame
CDH	Constricted double-heterojunction diode laser
CDH-LOC	Constricted double-heterojunction large-optical-cavity diode laser
CDM	Color-division multiplexing
CDMA	Code division multiple access
C/kT	Carrier-to-receiver noise density
CMRR	Common mode rejection ratio

Acronyms and Abbreviations

CNR	Carrier-to-noise ratio
CNS	Channel-narrow stripe diode laser
COAT	Coherent optical adaptive technique
COD	Catastrophic optical damage
CODEC	Coder-decoder
cpi	Characters per inch
cps	Characters per second
CPU	Communications processor unit
CPU	Central processing unit
Crank-TJS	Crank-type transverse-junction-stripe diode laser
CSC	Circuit switching center
CSP	Channel-substrate-planar diode laser
CSU	Circuit switching unit
CVD	Chemical vapor deposition process
CVPO	Chemical vapor phase oxidation process
CVSD	Continuously variable slope delta modulation
CW	Continuous wave
D-A	Digital-to-analog
dB	Decibel
dBa0	Noise power in dBa referred to or measured at 0TLP
dBa, dBrn adjusted	Weighted noise power in dB referred to -85 dBm
dBa(F1A)	Noise power measured by a set with F1A-receiver weighting
dBa(H1A)	Noise power measured by a set with H1A referred to 1 milliwatt
dBm0	Noise power in dBm referred to or measured at 0TLP
dBm0p	Noise power in dBm0 measured by a set with psophometric weighting
dBm(psoph)	Noise power in dBm measured by a set with psophometric weighting
dBr	Power difference in dB between any point and a reference point
DBR	Distributed Bragg-reflector diode laser
dBrn	Decibels above reference noise
dBrn(144-line)	Noise power, in dBrn, measured by a set with 144-line weighting
dBrnc	Noise power in dBrn, measured by a set with C-message weighting
dBrnc0	Noise power in dBrnc referred to or measured at 0TLP
dBrn(f1-f2)	Flat noise power in dBrn
dBW	Decibels referred to 1 watt
DCE	Data circuit-terminating equipment
DCPSK	Differentially coherent phase shift keying
DDD	Direct distance dialing

164

Decibel	dB
DEMUX	Demultiplex
DFB	Distributed-feedback diode laser
DFSK	Double frequency shift keying
DH	Double-heterojunction diode laser
Dibit	A group of two bits
DO	Design objective
DoD	Department of defense
DPSK	Differential phase shift keying
DSE	Data switching exchange
DSTE	Data subscriber terminal equipment
DTE	Data terminal equipment
DTMF	Dual-tone multifrequency signaling
E&M	Receive and transmit (signaling leads)
ECCM	Electronic counter-countermeasures
ECM	Electronic countermeasures
EDC	Electro-optic directional coupler switch
EFL	Equivalent focal length
EMI	Electromagnetic interference
EMS	Equilibrium mode simulator
ERP	Effective radiated power
FDM	Frequency-division multiplex
FEC	Forward error correction
FET	Field-effect transistor
FFL	Front focal length
FM	Frequency modulation
FO	Fiber optics
FOC	Fiber-optic communications
FOTS	Fiber-optic transmission system
FPIS	Forward propagation ionospheric scatter
FSK	Frequency shift keying
FWHM	Full-width half-maximum
FWHP	Full-width half-power
G/T	Antenna gain-to-noise temperature
GHz	Gigahertz
GMT	Greenwich Mean Time
HDLC	High-level data link control
Hz	Hertz
I/O	Input/output
ICW	Interrupted continuous wave
IFS	Ionospheric forward scatter
ILD	Injection laser diode
IOC	Integrated optical circuit
IR	Infrared

IRW	Inverted-rib-waveguide diode laser
IVPO	Inside vapor-phase oxidation technique
J	Joule
K	Kelvin
K	Coefficient of adsorption
k	Boltzmann's Constant
kHz	Kilohertz
km	Kilometer
LD	Laser diode
LEA	Longitudinally excited atmosphere laser
LED	Light-emitting diode
LNA	Launch numerical aperture
LOC	Large-optical-cavity diode laser
LOS	Line of sight
LP	Linearly polarized mode
LPC	Linear predictive coding
LPE	Liquid-phase epitaxy
M	Material dispersion parameter
Maser	Microwave amplification by stimulated emission of radiation
MBE	Molecular beam epitaxy
Mbits	Megabits
MCSP	Modified channel substrate diode laser
MCVD	Modified chemical vapor deposition process
MFSK	Multiple frequency shift keying
MHz	Megahertz
MIVPO	Modified inside vapor phase oxidation process
MOCVD	Metalorganic chemical vapor deposition
Modem	Modulator-demodulator
MS	Molecular stuffing process
MTTF	Mean time to failure
MUX	Multiplex
NA	Numerical aperture
NAM	Nonabsorbing-mirror laser
NEP	Noise-equivalent power
NF	Noise figure
nm	Nanometer
nmi	Nautical mile
NPR	Noise power ratio
NRI	Net ratio interface
NRZ	Non-return-to-zero
NRZ1	Non-return-to-zero, change on 1s
OCR	Optical character recognition
OMVPE	Organometallic vapor-phase epitaxy

OSDM	Optical space-division multiplexing
OTDR	Optical time-domain reflectometry
OVD	Optical video disc
OVPO	Outside vapor phase deposition process
P/AR	Peak-to-average ratio
PACVD	Plasma-activated chemical vapor deposition process
PAM	Pulse amplitude modulation
PCM	Pulse code modulation
PCW	Plano-convex waveguide laser
PD	Photodetector
PDM	Pulse duration modulation
PF	Packing fraction
PEP	Peak envelope power
PFM	Pulse frequency modulation
PM	Phase modulation
ppb	Parts per billion
ppm	Parts per million
PPM	Pulse-position modulation
PSK	phase shift keying
PTM	Pulse time modulation
pW	Picowatt
pWp	PW, psophometrically weighted
pWp0	Psophometrically weighted power in picowatts relative to 0TLP
QPSK	Quadrature phase shift keying
rad	Radian
RAPD	Reach-through avalanche photodiode
REED	Restricted edge-emitting diode
RF	Radio frequency
RFI	Radio frequency interference
rms	Root mean square
RZ	Return to zero
s	Second
(S+N)/N	Signal-plus-noise-to-noise ratio
SF	Single frequency
SI Units	International system of (metric) units
SID	Sudden ionospheric disturbance
SLD	Superluminescent diode
SML	Separated multiclad-layer diode laser
SNR	Signal-to-noise ratio
sr	Steradian
SSB-SC	Single-sideband suppressed carrier
SWR	Standing wave ratio
SX	Simplex signaling

Acronyms and Abbreviations

T_0	Threshold-current temperature coefficient
TDM	Time division multiplex
TDMA	Time division multiple access
TE	Transverse electric mode
TEA	Transverse-excited atmosphere laser
TEM	Transverse electromagnetic mode
TGM	Trunk group multiplexer
TH-LOC	Terraced heterostructure large-optical-cavity diode laser
TJS	Transverse-junction stripe diode laser
TM	Transverse magnetic mode
TS	Terraced substrate diode laser
UHF	Ultra high frequency
UV	Ultraviolet
VAD	Vapor phase axial deposition
VPE	Vapor-phase epitaxy
VSB	V-grooved substrate buried heterostructure diode laser
VSWR	Voltage standing wave ratio
VU	Volume unit
W	Watt
WDM	Wavelength division multiplex